Children of Revolution by Anna Louise Strong
2011 Prism Key Press / www.prismkeypress.com

# Children of Revolution
## Anna Louise Strong

# Contents

# I. Ten Boys on the Volga

IT all began with just ten boys. Boys of thirteen, fourteen, fifteen years, without fathers or mothers. Putoff, Michaef, Smirnof, and other names too hard to pronounce. They were thin, scrawny boys, small for their age and hungry. For two or three years they had not tasted sugar. Black, sour bread of rye was all they had, and sometimes thin potato soup. But even of this there was not enough for boys to grow on. Not since the Hungry Year, when their fathers and mothers died, and they were stranded with hundreds of thousands of other boys, along the great River Volga.

The River Volga is one of the mighty rivers of the world. Far in the north it rises, where even in the hottest summers the rains and mists keep cool the long white evenings. And slowly, muddily, spreading over the great plains of Russia, it wanders southward till it comes to the inland Caspian Sea by the fabled city of Astrakhan Here Tartar fishermen spread their nets in the scores of channels into which the great Volga breaks in delta formation, sluggish, hot, below the level of the ocean. Here the sun blazes down pitilessly from early spring until late autumn; waves of blinding heat strike up from the dazzling waters and the ancient city streets; and everyone who can travels northward to escape towards the pleasant hills of the northern and middle Volga.

On both sides of the great length of the Volga, and for hundreds of miles in every direction, are peasants of many old races, tilling the earth in very primitive fashion. They live together in crowded villages, five hundred, a thousand, sometimes five thousand houses; and from these villages they go forth in the morning to their fields. Some of the fields are near and some are distant. Each man has little pieces of land in many different places. For this was the way the land came down from their grandfathers; and their grandfathers also handed down the very old ways of farming.

Because their ways of farming are very old, there come years when the wheat and rye harvest fails. These are known as the Hungry Years. One year in every three is a Hungry Year along the Volga. It is not altogether the fault of the peasants. It is the fault of the earth as it moves around the sun, and the sun as it travels through space. Something about these movements of the sun and the earth, over hundreds of thousands of years, is slowly drying up the great plains between Europe and Asia, the plains where the Volga flows. Hundreds of thousands of years ago there were great inland lakes here and much water. Perhaps some tens of thousands of years in the future, even the river will be dried up. Meantime the peasants live along its banks, and do not know enough about farming to escape the Hungry Years, when the sun is hot and no rains come for the wheat. The worst of all the Hungry Years was 1921. That was the year when a million people died. For the land was parched with heat till it cracked open; and the grain shriveled and died before it reached the height even of a few inches. And the peasants poured forth over the land, fleeing in every direction from hunger; while those who stayed at home made bread of grass and of twigs and straw. They ate this till their stomachs felt full; but from food like this they grew swollen and died. All over the land, for a thousand miles to the north and south they were dying.

Then because there was not enough food for all, the government gathered the children into children's homes, and from all over the land, and from foreign lands also, as far as America, people sent food for the children. So after the Hungry Year was gone there were many children living, whose fathers and mothers were dead. Putoff, Michaef, Smirnof and the other boys were some of these famine children.

And now they were fourteen and fifteen years old. And the children's homes said to them: "We cannot forever keep on feeding you. We also are poor, and there are younger boys than yourselves without fathers and mothers, roaming the streets. Who

8

wants to start a working group and learn how to earn a living? We have land and houses up in the hills that we will give you; and a few tools and some food for a little while."

Many boys volunteered, for they were tired of sitting around the children's homes, where the food was very little and there was nothing to do, and where all winter long they must sit on the big stove, because they had no shoes to go outdoors. So ten were chosen to start, and with them went a leader, Yeremeef, who used to be a peasant. The houses of Cherumshan were in a beautiful place, with wooded ravines through which ran little streams of water. But they were old, with many broken windows and with no furniture in them, only the empty walls. Here the boys camped out in the empty rooms and began their fight for a living. Already they intended to get more than a living. For they had promises of help from America, and they made up their minds to build here, in the desolate Volga region, a modern American farm where all might live and eat, and where more and more homeless, hungry children might come to join their commune.

Already, when they came to the houses of Cherumshan, a carpenter named Fedotov had come before them and had made three narrow wooden bunks. The rest of the boys slept on the door; they had brought from the children's homes five burlap bags filled with straw and five thin blankets. Under these they huddled together, for the October nights were growing cold. Also Yeremeef had brought from town two lamps, ten bowls, ten spoons and two iron pots for cooking; and some black rye flour and potatoes and oil from sunflower seeds. This was all they had to begin living, except the horse and old wooden cart that took the things out to them.

The very first night they held a meeting. They had a supper of thin potato soup,--just potatoes boiled in water, with a little sunflower-seed oil to supply some fat,--and then they gathered around the lamp and discussed the future. They decided to get up at six every morning, and have breakfast at seven; then work from eight to twelve and from two to four, repairing houses, cooking, making shoes and clothes and dishes. "Who wants to be carpenter?" asked Yeremeef, and six boys raised their hands. "Who wants to be shoemaker?"--Three volunteered for this. The last boy said he wanted to look after the horse.

Early the next morning they started to work. But Fedotov, the carpenter, had only his own set of tools, and these were already old. A saw, a hammer, a hatchet, and a couple of planes. Not many boys could learn to be carpenters with that. Besides, there was no wood to work with. So some of the carpenters were put on the committee for cooking and the rest went out looking for wood.

They found it,--two broken-down shacks that no one could ever repair to live in. Their leader, Yeremeef, went back to town to get permission to tear down the shacks, for everything in the Cherumshan houses now belonged to the town, and to tear down even a shack one must get permission. This was soon done, and the boys set to work, taking the shacks apart and making the boards into tables and beds. It went very slowly, for none of them had ever been carpenters before. Besides, there were very few tools. But little by little, with Fedotov showing them how, rough wooden bunks for sleeping began to grow under their hands. Then tables, benches, and a stall for the horse.

After three weeks the boys sent a delegation to the town. "We are working very well now, but there is little food. But near at hand is an old mill which is not running. Give it to us and we will run it, and from the grain that we grind we will get bread." So the town gave them the mill, and also sent out four more children, Nasipaef and Lameku who wanted to be millers, and two girls, Nastasia and Katherine, to help cook and sew. These also were famine children who wanted to leave the children homes and learn to make a living.

About this time some old clothes from America came to Russia, to give to famine children. Down from Moscow to the new colony came fifty coats and trousers, and fifty pair of good American shoes. But only three pair of the shoes were big enough for the boys to wear; Putoa got one of these and it lasted for two years. He went bare foot most of the time to save his fine strong American shoes; he wore them only when the weather was cold and he must go on a long tramp to town. The rest of the shoes which did not fit them they sold, and bought leather in the near-by town of Kvalinsk. So at last their shoe-shop began working, and a shoemaker came out from town to show the boys how to be shoemakers. For this also they had few tools and only two or three boys could work at a time, waiting for each other to finish with the knives or the sewing machine. But steadily they worked away till all the fourteen children had shoes.

And now a great piece of luck came to the colony. Two days' journey away the Quakers from America were giving out food to children's homes. And the Quakers had heard of this group of boys which was starting to build a big farm colony of children, and promised to give them food. So Yeremeef went on the boat for a day's journey north, and then took the train for a day's journey west, till he came to the town where the Quakers gave out supplies of food. He brought back with him a whole car-load. Very wonderful food that they had not seen for years. Sugar and cocoa, and lard! There were also a hundred blankets, for starting a big colony. And soap--the first they had seen for many years! For ever since the Hungry Year they had been too poor to buy soap, and had scrubbed their dishes with sand, or ashes from the fire.

So now, because they had so much food, and blankets, they invited fifteen more boys and girls to join them, from the children s homes in the town. Seven girls and eight boys came out; they were all children who were reaching the age of fourteen and the town refused to support them any longer. But they brought with them from the children's homes an old sewing-machine. With this the girls set to work to make underwear and clothes and mattresses.

For the old clothes that came from America were of all shapes and sizes. Most of them did not fit the children. Some were too big, but most were too small, meant only for little children. These had to be pieced and they made very funny looking clothes. But funniest of all were some minister's suits that were given by some kind Preachers in America. They looked very queer on these thin, scrawny boys of the Volga, when they went out in the barn to tend the horse, or when they worked in the carpenter shop or the fields.

In the big carload of Quaker goods there came also some ticks for mattresses. But the boys and girls held an assembly and decided that these ticks were too good to use for mattresses. They were ever so much better than the old clothes from America. They decided to cut up the ticks and make shirts and skirts from them. But how should they get mattresses? For many weeks they did not have any; they found some straw and made a heap of it in their rooms and slept on that. But now that the young carpenters were improving in work and the wooden bunks were getting finished, everyone began to think of having mattresses.

All around these boys and girls, in many directions, were villages of peasants. And every peasant had a couple of burlap sacks, which he used to carry grain to market. But now, since the Hungry Year, these sacks were getting worn, full of holes, and could no longer be trusted to hold grain. The peasants also were hungry and had not seen sugar or lard or soap for many years. So the boys of the colony set out with the Quaker food, and traded sugar and lard and soap for old burlap bags. They got these bags very cheap, for the harvest was over, and the peasant would not need his bags very much for another year. By then he hoped he might buy some good ones. Out of these old torn burlap bags the girls made mattresses, patching the biggest holes. The smaller holes did not matter, for the bags were to be filled with straw, not grain. If a little straw spilled out on the bed-room floor, that did not hurt anyone.

As fast as beds were made and mattresses finished, more

and more boys and girls came to the colony. By February there were fifty-seven in all. And still the Quaker food lasted. The first supply of potatoes and hour and oil was long since gone, but they sold the precious sugar and lard and cocoa and with the money bought the cheaper potatoes and flour. But always they kept back a little sugar for themselves. It was the first sugar there had been for two or three years.

Soon the girls had made underwear for everyone in the colony, two suits apiece. Then they began making underwear to get ready for more boys and girls who would come in the spring. For up and down the Volga the colony began to be heard of,--this place where a group of boys and girls were learning to be carpenters, shoe-makers, millers and farmers. They knew that no more boys or girls would come to them in the winter; for famine children had no shoes and coats to travel far in the cold weather. But as soon as the spring should come, and the boats begin again to travel along the Volga, they knew there would be other children coming to the colony.

Meantime a problem arose about dishes! The ten bowls and ten spoons with which the colony began were no longer enough for fifty-seven children. They needed pails also for water, and pans for washing. In the houses of Cherumshan was a ruined blacksmith shop. This the boys repaired, and two of the boys decided to be blacksmiths. They had a man to teach them, the same man who was helping them to run the mill. Sheets of old iron they found in ruined houses, and of these they made pans. They also found large quantities of old tin cans, left over from the days when Americans sent cans of milk to feed children in the Hungry Year. In their blacksmith shop they took these cans, sawed off the tops and hammered down the edges till they would no longer cut, and thus made cups and bowls for soup. In the carpenter shop they carved pieces of wood into long spoons and ladies. These were the only dishes they had for the whole first year of the colony.

And now, with the spring of the year, came the time for

ploughing. Money had come from America to buy horses and tools. For already in the early autumn when the boys first moved out to the Cherumshan houses, they took the name of an American, John Reed, for their colony. He was a pioneer from Oregon who traveled many lands, helping men fight for freedom. He went to Mexico and wrote about the wrongs of the peons there before their revolution. And he came to Russia, in the days of its revolution, and died from typhus, and was buried by the great red walls of the Kremlin in the Red Square of Moscow. And because he was the American who helped the Russians in their darkest days of revolution, the colony named itself after him.

Also they had invited an American to be guardian of the colony, and to help it secure horses and ploughs and a tractor and grow to be a good American style of farm. So that it could be an example to the peasants in the whole region, showing them how to plant crops that would not fail in the dry summers. The first gift from America came at Christmas time--$100 for a Christmas celebration. Long and carefully in their meeting the boys and girls considered this great sum of money. What should they spend it for that the colony most needed?

"We don't need a celebration," said one of the older boys. "But we need horses. For spring will be coming and the time for ploughing and how shall we plough?" So they voted to send Yeremeef and two of the boys many days journey away to Uralsk, where they heard that horses were cheap. On freight cars they went and on foot, and they bought five horses. It was a large number, but the horses from Uralsk turned out to be not very good ones. One of them died before spring, and another they traded for a cow. When springtime came, they had three of the Uralsk horses and two others to begin ploughing.

They were already proud of their record that first winter. In the carpenter shop they had made more than fifty cots and fifteen tables; they had made twenty-five stools and twelve long benches, and fifteen troughs for feeding pigs or washing clothes. They had made two carts, very rough affairs with wheels bought

14

from the town, and with long poles laid across the axles and fastened together, and a rough framework arising from the poles. This is the kind of cart the Russian peasant uses; it is called a telega. They had also made many repairs to the mill and the houses, fitting windows into the broken holes until the place could be lived in. They also made the wooden parts of two ploughs and three harrows, to be ready for the spring ploughing. These were very rough ploughs, with only an upright piece of iron fastened in place by the young blacksmiths--but they were better than nothing.

In the blacksmith's shop they had made several dozen cups and bowls, ten pans, twenty-four pails, ten kitchen knives and six pruning saws for the orchard. They repaired the metal parts of ploughs and put iron teeth in the harrows. Now, as the spring approached, they began to make spades and mattocks. And when it came time to dig in the garden, there were already ten spades and twenty-five mattocks that the boys themselves had made.

Also the girls had not been idle. They had made over and over 200 pair of underwear. This was a large amount, but it was made from very old clothes and soon wore out. No sooner was one set of underwear finished than another must be worked on. They had made 52 mattresses out of burlap bags. They had made over 47 overcoats and 30 dresses and 30 boys' blouses and 20 pairs of trousers. All these things were very hard to make, for there was only one old sewing-machine, and the girls sat in line to take turns using it. Whatever could be done by hand was done that way. And after the clothes were done, they were already very old and wore out quickly.

Meantime the little mill had been working, and already three or four boys were becoming accomplished millers. It was not so easy to get customers for the mill, for it was a poor and small one, and would only work a few hours in every day. The power came from a mill-pond, which had not much water. All through the night and most of the day the water must gather in the mill-pond, till, perhaps, towards evening there would be enough water to run the mill for a few hours.

The peasants did not like to come to this mill, because it ran so poorly. So the children organized a committee to go out and advertise among the villages. "Come to our mill," they said, "for we also are peasant children, orphans of the famine. Some day we will have a fine school here, that your children also will want to go to. And even if our mill is very small, still you should help us. We are making our price for grinding lower than the better mills. We ask only three pounds from every forty that we grind, and the others take four or five."

As a result of this advertising, more peasants came to the mill. And Nazipaef and Lameku ground wheat and rye month after month, from February till August. Most of the grain was ground in the winter and early spring; after May there was little work. But in this time the young millers earned for the colony 3,240 pounds of rye Sour, and 1,328 pounds of wheat flour, and 1,600 pounds of millet for porridge. This was enough for almost three months' food for the colony.

So with the Quaker sugar and lard and cocoa and soap, and with the flour and millet from their mill they lived not badly, making their own furniture and shoes and clothes and dishes, till the snows ran off the hills and down through the ravines to the great Volga, which moved under its ice until it broke through the frozen bonds and began flowing free to the Caspian. Till the season of mud came and went, and the ploughing began.

# II. The First Ploughing

ALL through the long winter months on the Volga, the colony named for John Reed had been growing, buried deep from sight or transport in the hills of Cherumshan. The ten boys who began it had grown to fourteen, then thirty; they had made tables, beds, benches, stools; they had made shoes and underwear and mattresses; they had repaired and run the mill. But now, with the coming of spring there were already fifty-seven children ready to begin the real work of the colony --farming.

They had many more houses now than when they started. The first boys began in a large wooden house of fifteen rooms, set in a deep ravine with high wooded hills around it and a distant view of the great river. It was the biggest of all the Cherumshan houses and needed the least repair. They called it October House, in memory of the October Revolution, and also because they came to it in the month of October. Then during the winter they took also a barn, a mill and two broken shacks to make beds with.

There were only two or three warm winter coats in the colony, and four or five pair of winter felt boots. But, taking turns, they went scouting in the ravines around the houses, to see what they could find. Half a mile away on a large level place by a good stream of water lay a whole cluster of little houses, with even one building of brick. Most of these needed repair before they could move into them. But during the winter they made windows and fixed up these buildings also, and called it Center Town, because it was the most central spot among all the houses. Still another half mile away, up a little ravine that at first they had hardly noticed, they found a long wooden house of one story, with many tight rooms; near it was a pond and a stream and a bath-house. Into this place the girls of the colony moved as soon as warm weather made it unnecessary to crowd together in one big house for warmth. They called this May Day House, in memory of the May-Day holiday, on which they moved in.

The lands of the colony also had grown. The town of Kvalinsk had given them almost one hundred and sixty acres. But these were badly scattered, in the old peasant style. The largest field was ten miles away from their houses, and the other fields were three and five miles away in different directions. This was the way the peasants were used to farming their land for ages.

When the American guardian of the colony saw this land, she said that was no way to farm, and that there was no use getting a tractor from America to work on such tiny scattered pieces of ground. Wasn't there anywhere a big farm where the colony could have land all in one piece? Yes, they said, there was a place like this, a wonderful big estate right on the great river Volga, sixteen miles away from the Cherumshan houses. But it would take more horses and ploughs and children and money than the colony had to farm it. It would take at least five thousand dollars just for tools and horses and machinery. So the American Guardian went away across the ocean to raise the money.

Meantime the boys and girls of the colony worked on their scattered acres, planning them as well as they could. The nearest fields they decided to sow to sun-flower and millet, for these crops need much work and weeding, and the children would have to go to them often. But the big piece of land ten miles away was for wheat, which could be ploughed and sown by one group of boys camping out. But the best and richest of all the land was in the Cherumshan ravines right next to the houses. Only it was in many small places between the hills. Here it was decided to grow vegetables, potatoes, cabbages, tomatoes, watermelons and squash. For this land could be easily worked even by the girls; and if the summer proved dry, they could water it from the streams. There were twenty acres of this garden land.

When the snow was gone and the season of mud was over, the boys of the colony divided themselves into groups and set off to the distant fields to do their ploughing, with the home-made ploughs from their blacksmith shop. It was too far away to come back every evening, so they intended to camp out in the fields. As soon as they got there, two boys went hunting for poles and the others began scraping up straw and stubble and grass from the fields. They set up the poles in the form of a low tent with one side open and the other side sloping down to the ground. Then they covered the poles over thickly with straw and grass. Soon they had made for themselves a rough shelter that would keep out the rains and wind. They gathered more straw and covered the ground inside the shelter so that it would be soft and warm for sleeping. Into this place ten or twelve boys could crowd at night.

Outside the shelter they arranged a place for cooking. Two upright sticks with a third stick horizontally across them, and a pail hanging from this horizontal pole. That was the whole of their kitchen. Every few days when the horses came down from the Cherumshan houses, they brought enormous leaves of black rye bread, fifteen or twenty inches across, and a load of potatoes and cabbage, and a little oil or lard. The boys boiled i the potatoes and cabbage and fat in the pail to make soup, and ate it with the black bread. In the morning they had bread with "sweet tea." But the tea was not made of real tea, for that is very expensive and comes from far away China. So they took wheat grains and roasted them, and steeped them in boiling water, and added a little sugar and called it tea. Sometimes there was milk in it, but not often, for the cows were far away at the Cherumshan houses.

As long as they had this food they did not complain, for it was better than they had had since the Hungry Year. There was plenty of bread so that they did not feel empty, and there was the good sugar and lard from the Quakers. When the older girls cooked the bread, all went well, but sometimes when it came the turn of the younger girls to cook, the bread turned sour and heavy. Then the boys in the distant fields felt they had a right to complain, when a whole week's supply of bread came down at once, and it was all of it hard and sour.

Yet still they kept on ploughing steadily through the spring. They began on the very first day that ploughing was possible. But all around them the lands of the peasants lay bare and unploughed, for it was Easter Week and the peasants were celebrating. That also was an ancient custom of the peasants, to plough in connection with the times of church festivals. All Easter Week they were going to church, and on Easter Eve they stayed up all night in church and then went home to get drunk, as a celebration because Christ was risen. So it was two or three days after Easter before the peasants began in their fields.

But the boys of the John Reed Colony cared little about church festivals. They began in the fields as soon as the weather was good. They got free seed from the government because they were famine children. All day long they ploughed, working in two shifts. And the day after their seed was in the ground, there came a warm, soaking rain. It was almost the only good rain of the spring.

For after that rain the sun began to beat down with all its strength, and the ground grew hard and dry. The seed of the peasants went into this hard, dry soil, and did not grow. So another Hungry Year came to the Volga Valley, not so bad as the great Hungry year when a million people died, but still bad enough. But in the fields of the John Reed Colony there was grain. Not very much grain, for the dry weather hurt their harvest also. But they had three times as much as the peasants. So now a new motto spread among the peasants, that "God loves work more than Easter celebrations."

Seventy acres of wheat the boys planted, and forty acres of sun-flowers for oil, and twenty-five acres of millet for porridge. And when the harvest came they had thirty thousand pounds of wheat, and fifteen thousand pounds of millet, and over twenty thousand pounds of sun-flowers. But best of all was the garden. For when the summer turned hot and dry, and they knew that the crops in the fields would not be good, the boys and girls together turned their attention to the little streams that ran through the Cherumshan ravines. With spades and mattocks they dug little ditches, leading the waters gently from the streams down over all their garden. Thus they irrigated their vegetables, and when autumn came they had two hundred thousand pounds of potatoes and cucumbers and cabbage and squash. They had enough to trade for meat and bread and sugar. They had enough to feed for a whole year the fifty-seven who had planted it. And they felt very proud to think that they could make their own living.

But by this time there were many more children in the colony to eat the food. In the month of May, as soon as the big steamers began to move on the Volga, the children's homes of Saratov sent twenty-five children to the Houses of Cherumshan to work in the gardens. But it took them all summer before they learned really how to do this work. Then in the month of August, when it was time to begin gathering harvest, thirty more boys and girls came from Saratov to help in this work.

They came hungry and without shoes or underwear or coats. The children's homes of Saratov promised to send food for them, but they also were poor, with many hungry children. So all through the summer months these new children ate the food of the colony. By this time the Quaker food, of sugar and lard, was gone. In the very hardest months of work the children were most hungry.

Sores began to come on their bodies for lack of fats. And they had no soap to wash their clothes or their faces and hands.

This was the very hardest time for the whole colony, these months before harvest. Everywhere in the Volga Valley it was hard also for the peasants. Few of them had enough bread to last till harvest. Many were again eating bread of straw and grass. The children of Cherumshan were better off than many peasants. But now they no longer had enough even of the black bread. They began to ration it, giving each one just enough to keep from starving. Yet for ail that they went on ploughing and sowing and weeding the gardens. For they were famine children who had seen their parents die of hunger. They knew that only hard work lay between them and another Hungry Year.

But now, in the midsummer just before harvest, the American Guardian came back across the ocean, bringing money for horses and cows and tools, and to buy extra underwear and food. And by the fall of the year, Saratov also began sending money regularly to pay for extra food for the children. And their own harvest came, with its big stores of potatoes, and smaller amounts of wheat and millet and sun-flowers.

So they began to lay their plans now to take Alexeivka, the big farm on the Volga river, sixteen miles away. Here there were over a thousand acres, and great houses, enough for two hundred children. Enormous stables, for forty cows and horses. A brick factory that could turn out a million bricks a year. Big workshops for repairing farm machinery and for making tools. And a giant flour mill that once ground grain for a score of miles around. Even across the Volga on the ice in winter the peasants used to come to this mill, in the days before the war and revolution, when it belonged to a grand duke.

But now the whole place lay idle and empty. The grand duke himself had never lived here, for he spent his time in Paris drawing only the revenues, from his estate. And the German overseer who ran it for him departed early in the Great War. During the long years of revolution, and civil war and disorder, the peasants helped themselves to horses and grain and cows and tools, for they were hungry and their horses had been long since taken for the armies. Then when the Hungry Year came, and thousands of peasants fled across the land in all directions, they camped out in the houses of this big estate and tore up doors and floors for fuel. Till at last order came again in the land, and the new government sent guards to take possession, and afterwards some workmen to repair and farm it.

But there were hundreds of estates like this scattered all

over Russia, without money or workers enough to put them in order. The big farm at Alexeivka was far away from the center, and control was hard. One manager after another came, and some were thieves and some did not know their business, and even the good ones did not have enough money. So they took only the apples from the two hundred acres of orchard, but more than a thousand acres of farm land lay idle. This was the great estate that the children could have if they could farm it.

Already through the long, hard summer they had learned something of farming. And money had come from America to buy horses and tools. So they sent to the government at Saratov and said they were ready to start in Alexeivka. They wanted two hundred and sixty acres of land to begin with. They wanted the right to repair the Big House and the stables, and to use the old ploughs that were lying unused in the barns.

At the same time they chose a committee of boys to go with Yeremeef to Saratov and buy horses. For already it was seen that the bad harvest that summer would mean cheap horses, since many peasants were trying to sell. At the same time another committee went out into the villages, buying cows for the colony. And down in the big barn at Alexeivka they found many ploughs, unused for seven years and somewhat rusty. These they repaired in the blacksmith shop and began ploughing.

With the month of October came the first birthday of the colony, and they celebrated it in the Cherumshan Houses. In the October House they built a theater and held a big celebration. The president of the town soviet was invited and all the town officials and the heads of education. Delegates from other children's homes came also. Even from far away Volsk, the county town, came representatives. Nazipaef, the young miller, was elected chairman, and one after another the children made reports of the work of the year--of the harvest, and the carpenter shop, and the shoe shop, and the plans to take Alexeivka.

After this came feasting. There were little pigs in the

colony by this time, and one of these was killed and cooked. There were potatoes in plenty from their own garden. There was also one hundred pounds of honey. But there were no dishes. All night long the girls worked in the kitchen, peeling potatoes for the soup, for they had just one kitchen knife. All night they took turns, and all next morning and afternoon. So when dinner time came, the older ones were so tired that they could not keep order. So something happened before all the guests that made everyone ashamed.

It was the tradition of the colony that everyone should share equally, and that the committee in charge of food for the day should do the dividing. For this they had their self-government and organization. Two days before the festival I bought fifty cents worth of candy and gave it to the girls in the kitchen to serve for supper. But they locked it up instead. "Not till Saturday," they said, "when everyone is back from the fields." They did not wish the ones who were working far away to lose their share.

But now, on the day of celebration, while most of the colony was busy in the meeting, some of the smaller boys discovered the honey and made a rush for it. They spread great hunks of it on their bread and got it all over their faces. Then suddenly one of the older girls discovered them and ran to call the chairman. After that there was order, but much of the honey was gone, and everyone was ashamed that on the day of celebration, when guests were around, the smaller ones had shown themselves so piggish.

But this was soon forgotten for at the end of the celebration, Yeremeef arose and announced that the colony had been given as much land as they could plough in Alexeivka, and also the Big House by the river, and the big stable, and the Horseman's House and the old brick barracks, where a hundred soldiers used to sleep at harvest, when they got in the grain for the grand duke. More and more land we could hope for, as fast as we could use it. And the colony began its plans for taking

Alexeivka, and building there a big farm colony that should be known up and down the Volga, and where homeless children for hundreds of miles might come and learn farming.

# III. We Take Alexeivka

THE first boys who went to Alexeivka were a small detachment of twelve who ploughed the fields by day and slept in the barn on the cold October nights. They were camping out, just as they had done in one field after another; for the center of the colony was still in the Cherumshan Houses. The Big House in Alexeivka lay cold and empty on the edge of the river, with windows and stoves broken from seven years neglect. They must repair this house before they could live in it. Meantime, the important thing was the ploughing. For soon the land would freeze and the autumn ploughing would come to an end. And if the colony made a good record, they could count on more land in the spring.

So the twelve boys knew that on them depended the growth of the colony. Every morning and evening they worked, but they rested in the middle of the day. At first the noon-day rest was a long one, when the days were hot, and the morning and evenings light and cool. But as the days grew shorter and colder, the two shifts of boys began to follow each other more closely. The bread still came down to them from the Cherumshan Houses, and their soup they made as before in a big pail over an open fire.

It was here that I found them one day in late September, as I came up the river from Saratov. With me went a boy of eighteen from Moscow, to visit his brother in the colony. He took with him a suitcase full of books, which he had begged from many officials in the big city--books on farming, and bee-keeping and cows and horses, that he was taking down as a present to the colony.

Already the great Volga was running low in its muddy banks, for the summer had been hot and the year was growing late. The boats ran very irregularly, waiting for each other to pass the sand-banks. It was after one o'clock in the morning when we came to Alexeivka, and we hardly knew whether to get off or go on to the next stop, and make for Cherumshan. But some of the men on the dock at Alexeivka told us that a group of boys were already ploughing here, so we decided to get off and find them.

It was very dark along the road; we could see neither houses nor barns. After stumbling over the beach for half an hour we saw a light. Some of the workers on the big farm were still awake. We hailed them with a shout, and between the barking of many dogs they answered: "Yes, there are twelve boys here from John Reed Colony. They are camping in the big stables."

We turned in the other direction to hunt the stables. And again stumbled in the darkness through the uneven fields. At last we decided to untie our blankets and wait in the fields till dawn, which was not far off. But hardly had we spread the blankets on the ground when we heard the tramp of horses. Nearer and nearer they came in the darkness but we did not know who they were. Then we heard voices--they sounded like young voices. We shouted: "Do you know where the John Reed Colony is?" Back came the answer in high boyish treble: "We are John Reed!" Then they came nearer, and we saw that they were two boys who looked after the horses by night, giving them pasture and taking them down to the river for water, and watching them so that they should not wander away before morning. Quickly they showed us the way to the stables. There inside we found ten other boys and an older horseman, curled up on a pile of straw with a thin blanket wrapped around each. They brought out great armfuls of straw for us to sleep on and we lay down in the stable door with a thin moon shining on our blankets. It seemed hardly a moment till I was awakened by the rustle of boys going out to begin the morning ploughing. It was already red in the east, but the sun was not yet up.

About nine o'clock the first group of boys came back from ploughing, and the cooks had already made ready the thin tea and bread. After breakfast young Vanya, who had come with me from

Moscow bringing books, opened the big suitcase which he had carried all night in our wanderings, and brought out magazines and pamphlets. The boys fell on them, each picking out the subject that interested him most, and began to read until time for dinner, sitting in the shade of the stables. They had to knock away the flies with one hand and hold the book with the other.

Before dinner-time the committee of cooks approached me. "Can't we buy some fat in the village?" they said. "No one has come from Cherumshan for a week and we have had no fat for two days..." "What kind of fat?" I asked. "Oh, any kind, to put in the soup. Whatever is the cheapest. Butter or lard or sun-flower oil; we think the sun-flower oil is only a few kopeks a pound. And also we need some straw sandals to tie on our feet for the ploughing. The stubble in the fields is hard and at night the ground is cold." We bought the sun-flower oil, and also some milk for the soup, and a dozen pair of straw sandals at about five cents apiece.

So steadily the ploughing went on, by dawn and evening, and even by moonlight. A few days later the committee sent to Saratov to buy horses came back with four new ones, riding them bareback for a hundred miles along the river bank for several days. Then work speeded up in earnest, for there were only a few weeks left till the ground would freeze and the ploughing be over. And by the record these boys made would the future of the colony be judged.

And now came the cold rains of autumn beating down on the fields and on the thin coats and bare legs of the boys. It was not always possible to plough now, both because of the condition of the ground and because of the lack of warm coats. The ones who tried to work in the rain fell ill from colds and fevers. Yet still, between times, they went out to the fields. The ones who had sheepskin coats worked longest; then they loaned their coats to other boys, and they themselves crouched under the straw in the stables for warmth while the others ploughed. And when at last the winter fell, more than two hundred acres of land lay black and ready for the spring sowing, besides some fifty acres already sown to rye on the other side of Cherumshan. And there was a promise now to give the colony another two hundred acres or more in the spring because of the fine record the boys had made. They had done far more than any of the grown-up managers who had held the farm since the revolution. And all around the peasants were saying: "See how that colony works. It is sure to get ahead."

And now the ploughing group went back to the Cherumshan Houses, and a new group came down to take possession for the winter. Eleven boys and three girls were the first detachment. Four of the boys were young millers, who came to repair and open the big mill at Alexeivka. Three were carpenters, to repair the houses. Four were horsemen, to look after the horses; for it had been decided to keep all the horses for the winter in the big stables on the new farm. Three girls came also, Shubina, Infelina and Yershova, to wash and cook. A little later came Stesha and five more carpenters. By the time Alexeivka lay snow-bound there were 26 boys and 7 girls encamped here, to repair houses and mill, and look after the horses. The rest remained in Cherumshan till spring.

When the first group came down it was already the season of mud. The roads on the side hill were slippery and deep with mire and the horse that brought the tools could hardly travel. The boys and girls who had shoes took them off to save them, for they must wade at least ankle deep in mud and water all the way. From time to time on the side hills they stopped to help the horse or to hold the wagon from sliding. It took all day and far into the night to make the sixteen-mile journey. So they knew that no other horses would make the trip until the season of mud was over, and that meantime they must depend on themselves and the supplies of food and tools they had brought with them.

They camped out at first in the big stable, as the ploughing group had done. But the heavy rains were leaking badly through the roof, which had not been repaired for many years and was full of holes. Near by was the Horseman's House that had been given to the colony. But families of workmen from the big estate still lived in it and refused to move out till the season of mud was over. "For where can we go in all this mud?" they said.

So they turned to the big brick barracks where once a hundred soldiers slept in summer. This also was not built for winter use and had no heat. Its floor was of brick, now badly torn up in many places. But at one end of it, towards the river, there were two rooms with a stove and a wooden door and a basement beneath them. The biggest room was painted bright blue, and they called it the Blue Room. Into these rooms the boys and girls all moved together.

Here at least the roof was good and kept out the rains. But all the windows were broken and the cold winds blew in from the river. Far away to Kvalinsk they must go for glass, and twenty miles of mud lay between them. So they piled straw from the stables in the most sheltered corner (since the bunks they brought from Cherumshan were too cold), and all huddled together under their thin blankets, except when they were actively working. But day by day, as the cold grew more bitter, they comforted themselves with the thought that when at last the ground froze hard and firm, their horses could go to town and get glass for windows.

At last came the heavy freezing, and the snow. Then life that had been immovable, swamped by seas of mud, began to move again along the snow road. The workers moved out of the Horseman's House, taking with them the panes of glass that belonged to them. But the boys of the colony went to Kvalinsk by sleigh for glass, and began rapidly mending windows everywhere, in the Blue Room and in the Horseman's House.

For a month longer the girls lived in the Blue Room, while the boys had already moved to two rooms of the Horseman s House and were repairing the other two. But, although the windows had been repaired, the Blue Room was very cold. Beneath it was a large, open basement, into which the wind blew great drifts of snow. There was not glass enough to repair this basement; they merely stuffed the window-holes with straw, but the stronger winds blew this out. The cold came into the Blue Room through cracks in the floor and doors and through cracks in the walls around the windows, where the old barracks had settled for seven years into the ground.

In the corners of the Blue Room the potatoes froze and the walls were covered with frost. So the girls put their plank beds together in the middle of the room near the stove and all slept together as closely as possible under both blankets and straw. They took into the Blue Room one of the little pigs that might freeze to death in the stables; the boys had the other two piss in their rooms. All night long the little pig could not sleep, for he also was cold. He went walking under the beds making noises, and keeping the girls awake. One night he found the basket where the bread was kept and managed to open it and eat all the bread before morning. But it was still warmer here than in the stables, so the little pig did not freeze to death.

Colder and colder grew the winter. At last, in January when all the four rooms in the Horseman's House were tight and sound, the girls left the Blue Room and moved over all together to the same bunk house with the boys. The millers slept in one room, the horsemen in another, the carpenters in a third, and the girls in the fourth. Seven girls in one small room, and twenty-six boys in three rooms. They took with them three little pigs and a young calf.

But here also there were troubles. For the heat in the girls room came only from the kitchen. If they wanted to shut the door and be alone, they froze; but if they opened the door then the three pigs and the calf walked around their room all night and kept them awake. And even then there was not enough wood to keep warm always. Wood could be brought from the forest four miles away, and the boys went for it with horses. But the horses were cold and without enough grain to eat; and the boys had not coats to go round.

On bright, sunny days the girls would lend all their coats to the boys, and the sleighs would set out for the distant woods to bring in fuel. But on stormy days the coats of all the girls and boys together were not enough, and they dared not face the blizzards with so little protection. So when many stormy days came at once, the fuel gave out and the house grew cold. Then the girls would sit on the big kitchen oven, all seven together, dangling their feet and trying to absorb the last little bit of heat.

Coldest of all it was in the bath-house and laundry overlooking the frozen river. When the girls went there to wash clothes, they could look up from the steaming tub and see right through the cracks of the wall the ice of the river. The clothes froze there long before they dried, almost before they could be hung on the line. So quarrels arose between the boys and girls. For the girls said the boys got their clothes dirty too often and refused to wash them in the frozen laundry. Even their own clothes they washed very seldom. And the boys cried that the girls were selfish, and refused to chop wood for the girls' quarters.

But between these quarrels they knew that they must work together and help each other, if they were to live through the winter at all. Only three of the girls, Stesha, Garshina and Gudkova, had shoes, and kerchiefs for their heads. But they loaned these to all the other girls, whenever anyone had to go outdoors in the snow. By the time spring came, the three pair of shoes were quite worn out, and when the snow began to melt under the sun's rays, the girls went everywhere barefoot in the melting snow. But even when Stesha saw her shoes wearing out, she knew she could not refuse the other girls the chance to wear them. And however bad the quarrels grew between boys and girls, the girls never refused to give their coats when the boys went for fuel.

Meantime the work went on. The eight young millers with an older miller from the village to help, made a new drum for the mill and mended the silken sieve that was worn by many years of wear and began even in the autumn to run the mill. There were three mill stones, any one of which would grind 7,000 pounds of flour in an eight-hour day. But they had not money enough to repair more than one mill-stone, and this was quite enough. For not since the Hungry Year had there been enough grain in the whole district to give work to all the mill-stones of this giant mill. But now, from all the villages around the peasants came with their stores of grain, and the single mill-stone did all the grinding that was needed, piling up all the time three pounds from every forty for the food of the children's colony. And the John Reed Colony became known for many miles as the first folks that were able to open the big mill at Alexeivka after the civil war and revolution.

Meantime, during the long winter, the carpenters were also busy. They tore down some broken drying-sheds and from them made wooden bunks, just as they had done the first winter at Cherumshan. Fifty more boys and girls had come from Volsk to the Cherumshan Houses, so more beds were needed there also. But the longest, hardest job of the winter was the repairing of the Big House, that lay right on the river bank, with over thirty rooms, enough for a hundred children. Here were many windows to be made, and tables, and benches; and holes in the plastering to be patched, and walls to be calsomined. Almost to the end of the winter the carpenters worked, repairing the Big House.

One day when Yeremeef went on a trip to the town of Volsk, forty miles away across the snow, he met there a fisherman from Astrakhan, the fabled Tartar city from the southern end of the southern end of the great Volga River. And the fisherman said: "We have rented some fine summer houses not far from you on the Volga, for the use of the fishermen of Astrakhan next summer. And we need some furniture for our houses--tables, bureaus and stools."

Then Yeremeef answered: "Give us the order to make them in our carpenter shop." And he got the order with money paid in advance for material and some tools. So the carpenter shop worked also making tables and bureaus for sale; they made twenty tables and twenty-nine stools and fifty little cupboards and bureaus. And after they had paid for their lumber and paint, they had money enough from their work to buy a cow for the colony.

So, little by little, the colony grew in strength. But life still was hard. There was a teacher who came from the village to give lessons in reading and writing. But when she came there was no school-room, and the girls were lying in bed to keep warm, except for the two who had to get up to cook. So the teacher sat down in the room with her sheep-skin and her high felt boots, and gave them pencils and copy-books to write under the blankets. Even their heads they covered with the blanket, leaving only a little crack for light to work by◆ That was the way they studied that winter in Alexeivka.

But at last the springtime sun, shining longer and longer each day, began melting the snow. Down from the hills through little ravines it ran away to the Volga, leaving little channels of muddy brown. By the early days of April the snows all over the land were soft with water, and the roads were long lines of impassable mud. Soon the higher land was free and brown in the sunlight, while the waters that had poured from it lay in a great floor on top of the still frozen river, till under its weight the solid ice of the Volga began to break. Great cracks ran across it, great masses of ice broke loose and slipped sluggishly downwards, till under the blue skies of mid-April, the brown and swollen waters ran steadily and unbroken to the sea. And still the spring moved north, releasing yet newer floods of waters to pour downwards into the ever widening channel. At Alexeivka the river was two miles wide; the low farther shore could no longer be seen across the muddy waters. They rose to the very edge of the Big House and washed the door-steps. The river steamers appeared, going up and down the river, making again connections with the outer world that had been broken during the long winter.

With the close of April the ground was dry for ploughing, and all the old horses and the new ones that had been bought with money from America began moving forth over the new land given to the colony. Twenty horses there were now, and day after day they ploughed. By the time the roses bloomed in the tangled gardens of the old forgotten grand duke, more than four hundred acres lay sown, waiting for the summer rains and suns, to decide the future fate of John Reed Colony. And another hundred acres of good hay land, covered with lucerne that had not been cut for ten years, was given to the colony in recognition of its work. For nowhere, for a hundred miles along the Volga, was there such a great stretch of ploughed land as ours, belonging to a single organization.

The village, also, began to come to us for help. Poor and hungry and cold and ragged as the colony was, it was able to help others. For we had more horses and ploughs than anyone in the village. And the Self Help Committee asked us to plough over fifty acres for the widows and orphans of Alexeivka. In return for this they brought us an order for grinding 40,000 pounds of grain for all the village organizations. And since from every 40 pounds we got three pounds for our work of grinding, we got from our mill in May almost bread enough till harvest.

# IV. Our Ragamuffins

WE were sitting on the broken steps of the bath-house overlooking the River Volga. It was noon hour in midsummer and the smaller boys and girls were resting after a lunch of potato soup flavored with sunflower oil. Some twelve boys and girls were gathered round us, brown bodies sticking through the rags in places, faces shiny with sweat which even in this sheltered place ran freely from us, brown legs and arms caked with dirt from the fields, which would be washed off later by a swim in the river.

Two husky barefoot girls of fifteen, bearing between them a long iron trough full of torn, grayish underwear, came from time to time out of the bath-house and made their way, resting occasionally, to the river's edge, where they soaked and shook the laundry for one hundred children. For our water connections from the hills were not yet finished, and the old water pipes that once ran to the grand duke's estate have been torn up and blocked by the new waterworks which is to give decent drink to the whole village of Alexeivka. The brown river is our total water supply at present; girls on duty in the kitchen wear themselves out carrying water.

But from across the fields comes the toot of our English thresher, which we have repaired and which is threshing grain now for a dozen peasants, all of whom pay us three pounds from every forty. From it alone we have now two months' supply of bread. So, in spite of the heat and the smells of long-dead soapy water that arise from the half rotten planks of our bath-house, in spite of the thin potato soup for lunch, in spite of the swollen feet of the girls who have been carrying water, we feel a little proud of our record. We have even a tractor; we are considered very progressive. The peasants know of us for many miles around, and the famine children travel to us for hundreds of miles along the river.

A girl of fifteen lies next to me, face down across her arms, stirring in uneasy slumber as the flies chase themselves across her neck. She is tired from work in the fields, and even here in the midst of flies and chatter, in a crumpled posture on the bath-house steps she drops off to sleep. Once she stirs restlessly, as more than a usual number of flies cross her arms.

"Sleep, little daughter, sleep," says a soft voice beside me, in tones of tender mockery. I look up I to see a young half-naked imp of twelve years grinning down at the older girl and addressing her with this teasing diminutive. It is Feodor, whom I have not met before. He has come to us only this summer. He is one of the smaller boys who tends the pigs.

His deep brown shoulders stick out through ragged gray burlap that forms his shirt, and his rounded brown legs jut through festoons of rags that make his trousers, and he has a constant smile that is half impish, half angelic, and a bushy head of hair that makes him look like an African savage. We have just been discussing the new school and club house which we are repairing, from the old barracks where the girls froze last winter. "I will give recitations," says Feodor. "I can recite pieces about Ilych" (Lenin).

"Where did you learn them?" I ask, and he answers: "In Yaroslav," mentioning a city almost a thousand miles to the north. And then, cheerfully grinning, he begins to tell me the story of his life.

"Nobody sent me to the colony," he says. "I just came myself. I look after the pigs in the morning and evening, but now it's too hot. I reap wheat, too, where the thistles are thick and we have to cut by hand. My father and mother died in the Hungry Year, so I set out traveling with my two sisters. We went anywhere that looked good."

"Did many places look good?" I ask him, and he nods. "In

Samara and Saratov I worked in orchards. I went to Buzuluk and my sister stayed there in a children's home. They were smaller than I, and I found no place there. I went to Astrakhan and to Yaroslav and to Sezeran."

Astrakhan, a thousand miles to the south, and Yaroslav, a thousand miles to the north--to both of these towns Feodor had wandered since the Hungry Year and he had found them both good. "The people of Astrakhan are all fine folk; they all give you bread," he says to me contentedly. And "I learned to read and write in Yaroslav and to speak pieces about Lenin."

When winter-time drew near he had managed to drift back to the village where his father died. There a kindly peasant took him in. "He let me stay all winter and he also fed me," says Feodor, for food must be mentioned separately.

"But why did you leave him?" I ask. Feodor smiles at my ignorance. "But there was never any bread left in spring," he says. "I left for bread."

"Yet he fed you when he had not bread of his own till harvest?" I ask in surprise....Then brown, ragged Feodor stares at me in indignant amazement. "But he was a friend of mine," he cries, surprised at my standards of neighborly help.

The sun creeps slowly around the edge of the bath-house. The heat of its rays touch the legs of the sleeping girl; she stirs uneasily, turns and awakens, brushing away the flies. "Do you like it here in the colony?" I ask Feodor. "Not bad," he answers. "I am going to stay here."

"Do you like the food?" I ask, thinking of the hard, black bread and thin potato soup. Feodor pats his little stomach and smiles. "Not so bad," he says. "I'm full."

"But don't you want a shirt?" I ask, gazing at his festoons of rags. "They don't give you any," he answers with the philosophy of experience.

"Then why do you like it here?" I persist. "Here I have a place for myself and a roof at night," he says cheerfully. Then suddenly noticing that the others are listening, he grows shy, and hides his shyness with pertness. "My tongue is tired," he says, sticking it out for me to see. Then grinning he shoves Jakov into the foreground, demanding that he also tell his story.

The whole upper half of Jakov's small brown body is bare, but his trousers have fewer holes in, and he has, rather surprisingly, a cap on his head. He also is twelve years old and looks after the pigs with Feodor.

"I come from beyond Volsk," he says, "through one village to another village, I've been in the refuges since the Hungry Year....." "Your father and mother died then?" I ask, and he nods in a matter of fact way as if that is understood. "My brother went away, I don't know where. My aunt took us to the refuge--me and my three sisters.

"From the refuge they took the children far away to the places of bread. Me they took to Paskov. Then the peasants came for us; the ones who had no children of their own took us to work for them. After two years they said there was food again on the Volga and all of us who wished could go home. The peasant wanted to keep me, but I wrote down that I wanted to go home."

"Can you write?" I ask him. "Oh, no," says Jakov, "but a teacher came to see me and asked if I wanted to go home and I said yes, because the peasant made me work very hard; he made me cut hay and harrow and drive the wagon. Also I wanted to see my sisters again. So I came back again to the refuge at Volsk. Here it was better, for you could work a day and then just walk around for a week. In summer we went to the country and gathered apples. But I ran away because the feeding was bad. Here in the colony it is better than the refuge; there is more food, and it is not much work driving pigs. Besides, there is shoe-making here and I want to be a shoe-maker like my father."

Then Lushnof tells his story. Lushnof is the clown of the

colony. He can give imitations of the way Yeremeef talks to the idle girls--imitations that make even the ones he is making fun of laugh. Lushnof was born in the storied city of Tashkent, many days' journey away in the heart of Asia. His father was a railroad worker, who died in the Hungry Year, and his mother came back to her people on the Volga, and died there the year after.

"Then our aunt came to live with us, but she was a bad woman and beat us, and we didn't want her. I ran away at first, but then I went home and asked a neighbor, who was a good man, if he would move to our house and take care of us. He was an old revolutionist and after a while the Education Board gave him a paper that he should look after us, and they would give us food.

"But our wicked aunt went to the courts about it. For she wanted to live with us because we had a good house and a cow. There were four of us, one sister and three brothers. The old revolutionist went away at last because he was too busy with his work to bother with all those lawsuits that my aunt started, and she got the house. But we children would not live with her, and they gave the cow to my sister and put her in a children's home. The cow gives milk to all the children, but when my sister grows up and can leave the home, it will be her cow.

"My brothers also are in children's homes in the town. But I heard about this colony and asked to come here. I worked in the blacksmith shop; I want to learn about machines. Then when I know all about machines, I will get our house back, if it is still whole after my aunt has used it; and my brothers and sister will live together again and work in the town."

The girl who has been sleeping is now completely awake. She makes a teasing remark to Lushnof; he puts his foot on her shoulder and pushes her away. She jumps up with a laugh and stands threateningly above him. Lushnof carelessly sinks his teeth in an apple, biting without choice through the good spots and rotten spots. I see his teeth sink into a brown, rotten spot, but he eats it as cheerfully as a pig. He finishes it slowly, wormholes and all, saving just enough of the core to throw at the girl. It spatters on her blouse, adding one more spot to the many spots that darken her garments. She catches it expertly and hurls it back, hitting him on the forehead.

As Lushnof leans his head against my shoulder with a grin, I see that he is black and a little scaly under the hair. "Next time when you go for a swim you must wash your head," I remark. "There isn't any soap," he tells me.

Then a chorus of denunciation breaks out among the boys. "The girls have soap! They steal it from the laundry. They get soap to wash our clothes, but they put it in their boxes and keep it to wash with, and our clothes come back dirty. That is why the girls are always clean."....I change the subject by asking Lushnof if he still has the toothbrush and comb I gave him. Yes, he says, it is in his box. I started to ask why he never used it, but I thought better of it. The reason was clear enough. With no water nearer than the river, and no pockets in his ragged trousers to carry a toothbrush, and no handkerchief to put it in, how could he take a toothbrush to the river for the purpose of extraordinary cleanliness, and then walk all the way back to the house before going to work?

And now Lushnof begins telling about the civil war in his town. "We ran and hid in the cellar. Not our cellar for we didn't have any. But there was a high school near us with a fine big cellar and everyone went and hid there while the shooting went on. It was crammed full of people and we stayed there for two days. I saw a woman run out of her house with a baby in her arms and try to come to the cellar, and just then some soldiers on horses came round the corner fast, and before they could stop they ran right over the woman and killed her and the baby. We had bread with us in the cellar and there was a big barrel of water. But it was so crowded that you could not breathe. After two days the shooting stopped and we all came out.

"After that there was the Soviet Power in our town. Some

41

of the people were for the Whites and some for the Reds. Then the bandit Pop off came up the Volga and gathered all the Communists in our town and killed them all. But the Red Army came and drove Pop off away again. Some said Pop off was a big bandit with thirty thousand men, but when he came nearer they said he had ten thousand men, and when he came into the town they said it was one thousand men, or maybe only five hundred."

And now Jakov grows excited with the war tales and begins chanting the deeds of the Red Guards, while his brown body sways back and forth in hero-worship. He half tells, half chants how the town of Volsk was "full of bourgeoisie, with fine officers' clothes from England.... "But the Reds came up the river, and they were dirty, dirty, dirty!........ and they had no shirts! .... and they had only broken shoes!.......... and they were hungry!......... but for all that they drove out the Whites, with the stylish clothes from England ! ... And they opened up homes for the children,, who before were sleeping in the streets.

Jakov smiles down in half-conscious approval of his own dirty, ragged, brown body as he chants his tale of the men who were just like him--but for all that they took the city and the land.

# V. The Story of Stesha

STESHA is seventeen years old now. She has three times been elected secretary of the Children's Organization in the Colony; she is known to all as one of our most responsible girls. She came to the Cherumshan Houses in the second month of the colony, and helped organize the sewing of mattresses. Because she was so dependable, they sent her also with the pioneer group that took Alexeivka the next winter.

Stesha was one of three girls who had shoes and a winter coat. But she loaned her shoes to ail the other girls when they had to go outdoors in the snow. And she often loaned her coat to the boys when they went to cut wood from the distant forest. Always she thought of the good of the colony before her own. So her shoes wore out and her coat wore out, and in midsummer she fell sick from malaria, and they sent her up to the Cherumshan Houses to rest.

In the Cherumshan Houses she was now the only responsible older girl. She organized the drying of pears and other fruit from the old orchard, so that there would be better food in the winter for the whole colony. She organized the special baking of white bread for the sick children. All the life at Cherumshan she tried to organize.

At last she was well enough to go back to Alexeivka. She wanted very much to go, for I was staying there, her friend from America, and she wanted to be with me for the two weeks before I must go away again. I promised to drive her down in the wagon the very next day. But then someone of the girls said: "But who will manage the drying of fruit?"

Stesha looked at me hard for a moment and her eyes filled with tears. She got up suddenly and left the room; I followed and found her weeping. "I want so much to go down with you," she said, "but it is true, I cannot go. Or else there will be only bread and potatoes and kasha all winter to eat. But if I stay here now and organize the work, we will all have stewed fruit on Sundays."

So Stesha stayed in the Cherumshan Houses all through the autumn months, drying fruit for the colony's winter food. But before I left her, she told me the story of her life, sitting on the covered steps of one of the Cherumshan Houses, while the first beating rains of late August poured down.

Far away in the South Ukraine live Stesha's two brothers and her aunt. But she does not remember any of them. For her father came to the town of Saratov when Stesha was very young. There he died and her mother also, and Stesha was put in an orphan asylum long ago in the days of the czar and when she was only six years old.

The very first thing she remembers is how she came to the asylum and they put on her a long blue dress reaching her ankles and a long white apron and a bib at meals. Stesha cried when they did this, for ordinary children wore short dresses. And now from her long blue dress everyone would know she was an orphan, kept by charity in an orphan asylum. All the orphans hated those long clothes.

But the worst thing Stesha remembers were the terrible long prayers on Sunday. The director of the asylum was a priest and so was his first assistant. He intended that orphan children should be specially religious. So every Sunday morning at four o'clock, Stesha and the other girls went to church without any food. They stood there on the hard, cold stone floor, looking very straight ahead of them at the priest and the candles and the swinging golden censer. For hours and hours they stood, and dared not look to right or left. If any child so much as glanced aside, the director would scold her afterwards. Stesha wanted very much to be a good girl, and she wept when she was scolded. But it was hard to stand still, looking straight ahead, all Sunday morning.

Sometimes in the middle of the morning was a little stop in the church service, and the children could go outdoors and talk with each other; but it was not long enough to go home. Then the service would begin again and last till noon. After that Stesha went back to the asylum and they had house prayers. There was a special prayer room in the asylum for prayers every morning and evening. One of the children read the prayers and the others stood looking straight ahead. Nobody sits down in a Russian church.

Then the children went to the dining room, and standing around the table, they sang a hymn asking Cod to give them food. After this they ate, and they rose and thanked Cod for the meal. Then they could go to their rooms or walk in the courtyard till four o'clock. At four o'clock came evening service in the church, standing again, looking straight ahead, till seven o'clock or later. If there was a big holiday, the church lasted long into the night.

"Nothing in all my life was so dreadful as praying to God," Stesha told me. "We used to hide under the beds to get away from it." This was what religion meant in the orphan asylums of old Russia.

Week days were not so bad. Stesha rose at six in the morning, washed and went to the prayer room for an hour of morning prayers. Then came breakfast and at nine o'clock she began work in the summer or school in the winter. The older girls did beautiful embroidery; they sat in their room all day long doing it, and the asylum sold it for them, and put the money in a bank for them to have when they grew up and got married. After lunch the children were allowed to go out in the courtyard, but never in the street. Life was like a prison without any freedom. But the children were so used to it that they hardly minded it. Then in the evening they stood for an hour at evening prayers in the prayer-room, and went to bed.

Stesha remembers very clearly the time when the czar fell, though she was only nine years old. The old priest told it to the children in a frightened whisper. He said they must pray and pray very hard for the czar to come back, or nobody knew what terrible things might come to pass in the world without the czar. There might be earthquakes and fires and floods and the end of the world! Stesha was terrified and she wept every night and prayed for the czar to come back.

Then came the real revolution. There was shooting in the streets; the children could hear it. The frightened old priest gathered them all together in the cellar and there they stayed all day in terror, praying to God and weeping. At night the shooting stopped and they came upstairs to sleep. But they could look from the windows and see the red glow of fires where buildings were burning. They knew that this was the end of the world that the priest told them about. So they wept and prayed God not to send them to hell. Next morning they ran to the cellar again when the shooting began. This they did for three days and nights. Once a bullet came through the upstairs of their house.

But for all that the end of the world did not come. On the third day the shooting stopped, and everything was quiet again. Only now the director was even more terrified for he said there was a new government of Bolsheviks and they were sure to do something bad. And after a week there came some men from the new government to the orphan asylum and took the director away. At this the children cried again, for they were used to the director and he had filled them with such fear that they did not know whether the end of the world might not after all be coming.

And then a tall young Bolshevik stepped over to the children, and put his hand on the shoulder of one of them. "Don't cry, kids," he said, "we'll send you a better director, and you can wear short dresses and not pray to God any more." After this they all stopped crying.

And sure enough, a new director came, and gave them all short dresses. And they did not have to go any more to church. They were just like other children. All at once there was freedom; everyone could do as she pleased. Only no one knew how to act,

46

for they had been used all their lives to obeying orders, and doing only what they were told, and believing whatever was said to them.

So now at first there was very much disorder. Every girl got up in the morning when she wished, or stayed in bed if she preferred. Sometimes the breakfast was ready and sometimes it wasn't. The girls who had always helped in the kitchen and in the bed-making suddenly decided that they wouldn't do it any more. And the girls who had always been most orderly, began throwing their clothes about, just to see if they could do it. And the whole house got so dirty that no one could keep it clean. Even the servants and teachers couldn't clean up as fast as the girls got it dirty.

But after a little while the girls began to say to each other that this was no way to live. Then some organizers came from the Education Board and called a meeting of the children. And they said: "We are going to organize self-government. You must choose your own Children's Committee to govern the life of the home, and your own Children's Court to keep order. And if any girl does anything wrong, she must be brought before the Children's Court and punished. For you cannot all live together in comfort without order."

So now for a time all went well. But now there began to be more and more children in the children's homes, for everywhere in the land was fighting, and hundreds of thousands of people dying of war and diseases. And the children's homes in the city of Saratov got very crowded. At first they took houses of rich people who had run away from the Revolution, and put the children in these new houses. But there were not enough of these, because the fighting had destroyed many houses. Besides, the cost of food in the city kept going higher and higher.

When at last there were twenty-one children's homes in that one city, they began sending them out into the country, where they could live more cheaply in the big estates of the former nobles. So Stesha was sent away to a fine farm seventy miles away across the River Volga.

And now came the Hungry Year! But, strange to say, Stesha was not hungry. For a thousand miles around people were dying of famine, and the farm where she was staying was in the very center of the worst part of it. But it was an irrigated farm, far away in the hills and it had fine gardens. There was plenty of food on this farm, when everywhere else people died of hunger. The children on this farm would not believe that there was a Hungry Year.

So they took the children on an excursion to see the terrible things that were happening in the land. Stesha went to the town of Kvalinsk and she saw people dropping in the streets from hunger. She saw them shoveled into a great grave, a hundred or more people together; for there was no time or strength to bury so many dead separately.

After the Hungry Year the children's home where Stesha lived was closed, and she came to John Reed Colony. Already there were fourteen boys and girls who had been making wooden bunks and tables. But now Stesha, who knew much about sewing, began to help organize the sewing of mattresses. She took two worn sacks of burlap, that the peasants used to carry grain to market, and put them together for mattresses filled with straw. The girls also made two sets of underwear all around, and then they heard that there would be more children in the spring, so they began making underwear for them also.

"Life was good in the colony that first winter," says Stesha. "For there was only a small number and we were friendly like a family. The fighting between the boys and the girls had not started. For there was plenty of soap from the Quakers and the boys did not get their clothes so dirty in winter. So they did not blame the girls for dirty clothes, and the girls did not blame them either. Also we had sugar to eat, the first time in two years, and cocoa and fat. Until spring we had plenty to eat, and life was good. But then Saratov sent us too many children and no food. And we were hungry again and without soap for washing. And the boys and girls began to hate each other. Because the undershirts were not enough, and the boys would not give them to be washed till they were too dirty, and there was no soap, so the shirts wore out with much rubbing. And after every washing there were shirts that had turned to a heap of rags. Then the boys called out: 'Give us a shirt,' to the laundry girls; and the girls ran away and wept, because they had no shirts to give.

"It is easy to organize order when there is only a little family. But when there is a whole commune of a hundred children, then it is very hard. When you are all in a few rooms together, you can see every moment what happens to the sheets and the shirts and the lamps. Then people take care of things because others see them. But to make order, and to have people take care of things that belong to a big commune is much harder. I think it is the hardest thing there is. But also the most important. For if we cannot live together in peace and order in a commune of one hundred children, how will we ever build the Great Commune, of all the people in the Soviet Union, and some day of all the people in the world?"

This is the problem that worries Stesha, who is secretary of our Children's Committee. When there were only twenty-two boys and seven girls, they already elected the Orkom (the Committee of the Organization), and the Children's Court and the Committee on Civic Instruction. For a whole year Stesha was secretary of the Orkom, the highest committee in the colony, and Nazipaef, the young miller, was president. Then Vera said: "Why should those two have all the chance to learn how to run public affairs? We also want to take our turns in offices."

The children's assembly elected Vera then, and gave her a chance to be secretary. But after a month, they didn't like the way she worked, and they tried someone else. This time it failed also, so again they elected Nazipaef and Stesha for another year. For Nazipaef also did his work well and the boys listened to his advice.

It was not till after I left Stesha that I learned the tragedy that has come to her personally. For she talked so much of the life of the colony that she quite forgot to mention her own life at all. But down in Alexeivka they told me how in the long cold months of winter, crowded all together under their heaped up straw and blankets in a single room, Stesha had got trachoma from another girl who had it. So she must leave the colony and go for a long time to a special hospital in Saratov where her eyes can be cared for. And her heart, too, is not very good. The doctors say it comes from nerves, because she has worried so much over the disorder in the colony, trying to make things better. He says if she has good food and rest in a place where she does not feel responsible, her heart will be soon well again. But Stesha is one of those girls who will always feel responsible for others. Not till she sees the whole colony at peace and happy will she herself be able to be happy and at peace.

# VI. Odintsof, the Faithful Horseman

CONSTANTIN ODINTSOF has two pair of pants! And a pair of sandals! And a new shirt! The boys all gathered round him in his glory when he came back to Alexeivka at the end of the summer, and showed what he had earned as the result of his summer's work. But they said no word against all his new clothes, for he has earned this summer much money for the colony also, driving one of our horses every day for the fishermen of Astrakhan in their summer resort in the hills.

For Odintsof is our faithful horseman, and we trusted him to work by himself far away from control, and to take care of one of the horses of the colony. And the Astrakhan fishermen trusted him also, to take their trunks and hand-bags from the boat to their summer villa. And when they drove with him to market, they gave him tips to buy cigarettes. But Odintsof saved the money and bought himself all his new clothes. And now he is back in our fields, harvesting the lucerne.

He came to us a year ago at threshing time from the children's homes of Saratov, where he had spent four years since his father died. But long ago, young Odintsof came from the great plains of Siberia, where the farming land runs flat and fertile for more miles than anywhere else on earth. His father had plenty of land and two good horses, and ten children, four boys and six girls.

But there came the great war of the czar with the Germans. And the oldest brother, Vassili, went away to fight. He was taken prisoner and sent to Austria, and for a time he suffered very much in the, prison camps. But then he was given to an Austrian peasant to work on the land in Austria. His peasant boss was kind and had a fine daughter, so Vassili married the daughter of his boss and stayed in Austria to farm.

The war was lucky for Vassili, who fought and was taken prisoner. But it was very unlucky for the Odintsof family that stayed at home. Far out on those great plains of Siberia, the greatest plains of the world, where anyone might think himself safe, the Revolution and the Civil War found them. Cossacks came into the village and rode into the yards of all the houses. In the Odintsof stable they found two fine horses.

"Here are some good ones," shouted the Cossack leader, and began at once to saddle Odintsof's horses and take them away.

But the old farmer Odintsof came out and protested. "I have a large family and I need two horses to plough with." The Cossacks only laughed. But when he laid his hands on the horses' harness and tried to hold them, the Cossacks struck him in the face and kicked him. "Clear out, or you will be shot," they shouted. And as he started to move away, they ordered him, "Give us some oats."

So old man Odintsof gave his oats to the Cossacks, and then they killed his chickens and made his wife cook them. Last of all they took his oldest boy, that was left now in place of Vassili, to fight in their army. But the boy hated the Cossacks and ran away from their army and came home.

Now there was even greater danger for the Odintsof family. For the Cossacks considered his oldest son a deserter. So whenever the Cossack White Guards came through the village, their sympathizers would tell them: "Old man Odintsof has a son who deserted from you."

So they came at night to Odintsof's house seeking his son. "Tell us, old man, where he is," they demanded. Now the son was hidden in a potato cellar close by the barn, and the trap door that led down to it was covered with a heap of manure. So old man Odintsof said he did not know where his son was. But the White Guards did not believe him.

So they beat him on the head and took him out to a pond in the snow and made him cut a hole in the ice. "We will throw you in, old man," they said, "unless you give up your son." But still he refused to tell, so they threw him in, and when he was almost drowned they dragged him out again, saying again: "Tell us where is your son." This they did many times, till he was covered with ice. But he endured everything and at last they believed truly he did not know. And they let him stagger back to his house and take off his wet icy clothes and warm himself again. And all the time the son lay safe in the potato cellar under the heap of manure. After that the son did not trust the villagers any more and went away to join the Red Army.

After this came the Red Guards also to the village. But they didn't take any horses, because there weren't any left, and besides they had captured all they needed from the enemy. They cried: "Clear out of the village. There's going to be shooting." So old man Odintsof and his family took bread and water in a hand-cart and went into the forest where they sat all day and all night while the shooting went on in the village.

Many, many times this happened. Constantine does not remember how many. White Guards and Red Guards and White Guards again were always fighting in the village. But after many years the fighting stopped and life was good again and quiet. Only now the Odintsof family was beggared; they had nothing left, neither horse nor cow. So the mother and her oldest sons stayed on the land to hold it and try to farm it; but old man Odintsof took his five youngest children and set off for the town he had come from years before, far away in the old Ukraine. He did not know that there was even worse fighting there than in Siberia, and that all his old relatives were also ruined.

Old man Odintsof died of typhus in Saratov, and his five children were put in the children's homes. After this came the Hungry Year, and word came that the children should all be sent to places of bread. The two Odintsof boys were sent to Novgorod and the three Odintsof girls went to Voronezh, and since these places are hundreds of miles apart, Constantine has never seen his sisters since or heard from them. For nobody in the Odintsof family knew how to read and write.

Young Odintsof was twelve years old now; his brother was still younger. They put him with hundreds of other children in great freight cars and started them on the way to Novgorod. But the railroads were still badly broken after all the fighting, and full of cars of food going to feed the hungry. So the cars full of children moved slowly. Sometimes they would stop for many days at a station. The weather grew colder and colder till at last winter arrived. And still the car full of children was traveling on its way. Then the littlest children began to die; some of them froze to death and some died of hunger, but most of them died from different diseases that came because they were so weak with cold.

When children died on the train, they put the bodies in a separate car. Every morning they went through the train and took out the ones that died in the night and put them in this car. And then, when they came to a convenient place, they dug a great pit and buried together all the frozen stiff little bodies. And so at last they came to Novgorod where was a place of food. But here too it was very cold; there was no heat in the house. And Novgorod is very far north where the sun shines feebly in winter and the cold, damp winds come in from the Baltic Sea. The children had only thin shirts and trousers without any underwear or shoes. All they could do in the winter was to lie on the beds and shiver under thin blankets. But at any rate there was food and they did not die of hunger.

After two years the Hungry Time was over in Saratov and the children were told they could come back. Another freight train was filled with children and started on the homeward journey. But on the way young Odintsof fell sick with typhus. So they took him off at a station and put him in a hospital.

At last Odintsof got well and they let him out of the

hospital and gave him a paper that he could go. But he could not read or write and he thought it was a paper to take him to Saratov. So he went to the station and tried to get on a train. But the station police stopped him and took him to headquarters and asked him what he was doing. He showed his paper to the man in charge, and the man read it through and said: "Well, what about it?"

"It is a paper to send me to Saratov," said Odintsof. "Oh, no, it isn't," answered the man. "It is just a paper that says you can leave the hospital." Then Odintsof found that there were no papers about him anywhere and no arrangement had been made to send him back to Saratov. But he told his story and the police believed him and put him on a train with a proper pass to Saratov.

All this took much time, and when at last Constantine Odintsof reached Saratov, he found that his brother had been sent home to Siberia just the week before. He also wanted to go, but they said: "You must wait till we have a whole party going." So he waited for two years, but they never had a party. All this time he lived in children's homes in Saratov.

At last some boys in the home heard of the John Reed Colony, where they could begin to work for their own living and learn different trades in workshops. Five of them went to the colony and wrote back that they liked it. So Odintsof also decided to go. And when Sergei, the shoemaker, came down to the big city, he took Odintsof back with him. Odintsof wanted to be a shoemaker, but Sergei said the workshop was full. He had only enough tools for six, and he was already trying to teach twelve. So Odintsof went to work in the stables caring for the horses.

"I have been here a whole year," says Odintsof, "and Yeremeef has not once scolded me. Everyone else he scolds, but he says I am faithful and look after the horses well. He would not send any other boy away for a whole summer with a horse, to work for the colony. But me he sent, because he knows I take good care and will not spoil the horses."

For the fishermen of Astrakhan came north in the summer to find a Rest House, during their vacations, from the heat. It was they who ordered dozens of tables and bureaus and stools to be made in our workshops. They also asked if we could give them a horse and driver and milk for the whole summer. They paid some money ahead, and with this Yeremeef bought three cows, paying for them partly with this money, and partly later out of the milk. At the end of the summer the cows would belong to the colony.

So Odintsof went to spend the summer in the Rest House with the fishermen of Astrakhan. Two other boys went with him, and one girl named Claudia, to milk the cow. Claudia lived in the house with the manager, and brought him his dinner, and helped take care of his little boy. The manager's wife wanted to take her back to Astrakhan, but Claudia refused, for she did not want to leave the colony. But she got from the manager's wife a nice dress and a pair of sandals; while the colony got from the milk enough money to pay for all the cows.

The other two boys took the cows out to pasture and fed them. But Odintsof thought he had no work at all. He only had to drive to town twice a day, sitting comfortably in the wagon, and bringing back food for the Rest House. Often when the fishermen went with him, they gave him a few kopeks to buy cigarettes or sunflower seeds. Sometimes he bought sunflower seeds for all the boys and for Claudia, because they had no one to give them money. But most of the money he saved to buy clothes with.

He bought one pair of trousers second-hand but almost new for seventy cents, and a shirt for sixty eight cents, and a pair of leather sandals for a dollar and twenty cents, for leather is scarce and expensive. Then he got a best pair of trousers for one dollar. They were all of cotton goods, but firm and new.

"It was a very fine summer," says Odintsof. "A pine woods is the finest woods there is. The fishermen's wives who came cried when they had to go back to Astrakhan, for it is very hot there. We also had nuts and fruits. But I got tired of it, for I like to keep busy and not to sit around on the hills. As soon as I came back to the colony, I went right to loading wheat, and now that is finished and I am loading lucerne. It is a good thing to see at the end of the day the piles of work you have done.

"Some time soon I want to go back to Siberia, to find my brothers. I wanted to go in the spring, but they would not give me a ticket. And now it is too late. For if I should go in the autumn when all the work is over, then my brother would look at me and say: 'Why do you come now to eat when you did not come to work?' So I must eat here for the winter where I have worked and made a harvest. And next spring perhaps my sisters will come back from Voronezh and we will all go home together.

"I want to study here and learn to be a shoemaker. I told Yeremeef I would not be a horseman another winter. I like horses, but you cannot study when you are horseman. All last winter whenever I thought I was going to study, the boys would come and call: 'Odintsof, get the horses ready; we must go for wood!' This winter someone else can be horseman, and I will learn a trade. It is not bad to be horseman just for a year, when it is your turn."

A sudden glow came into the face of Odintsof. "Besides," he said, "there will be an orchestra this winter, and they have promised to let me learn to play."

# VII. Marin, the Tractorist

PETER MAIN is the boy who drives our tractor. Three other boys have learned to drive it, but Marin has it most of the time. He never gets tired of machinery. He can make the tractor plough eight acres of land in an eight-hour day--land that has grown hard and tangled from many years' neglect.

When Marin comes home from his eight hours with the tractor, he looks around for other machinery to work with. He hunts through the mill to see if they are getting it properly repaired for autumn. He hangs around the thresher to see if the belts are getting loose. If he can't find any machines to handle either in the fields or the mill, he comes around to repair my alarm clock which is always getting out of order. Or he fixes up some of the cheap lamps around the place.

Marin comes naturally by his love of machinery. His father was a mechanic who moved to the tiny village of Veshnik on the other side of the Volga, when Peter was two years old. The older Marin helped run a mill in the winter, and in summer he repaired ploughs and wagons. He joined the Red Guards at the very beginning of the Revolution and marched away to fight the White Guards. He never came back from the civil war: he was killed in some city somewhere. Peter does not know where; he only knows the word came back to him slowly that his father died for the Revolution. It is almost the first thing that Peter remembers.

For the memory of Peter Marin's childhood is very dim. He remembers first a horrible headache that seemed to last forever. After that he was in bed for many months. This was when he had typhus, the terrible disease that comes from the bite of a sick louse, and that raged over Russia together with hunger and cholera in the days of the Revolution. Peter was sick for six months with typhus, and when at last he could work again, the memory of the past was almost gone. He remembered his father vaguely, but he cannot remember anything his father ever did-- except one thing: it is burned into his brain proudly that his father died for the Revolution. Everything else was blurred out by the typhus.

So the first thing Marin remembers, after the typhus was over, is one day when he was twelve years old and the Red Army came through the village. There was one long ragged road that ran between the houses, and at one place it opened out into a wider irregular space around a church. The soldiers stopped in this open space and Peter, with other boys, came running to look at them. They were fine looking soldiers now, not like the ragged, dirty troops that first conquered the Volga, for it was the second or third year of the civil war, and they had uniforms and shoes now, and a regular Red Army, with Trotsky at its head. They were driving back the last of the White Guards and clearing the land of bandits.

"Hello, children," cried the soldiers in the open square of the village. "Who wants to come with us?" And twelve-year-old Marin cried out: "My father died for the Revolution. I want to go." Then a big soldier hit him on the shoulder and said: "Come along, kid."

So Marin went with the army. He did not even go back home to tell his mother or to get his things. For he hadn't any things to get. The shirt and trousers of worn white linen, and a tattered sheepskin that belonged to his father were all his belongings and these he already had on his back. And what was the use of telling his mother when she was sure to object? So Marin's little brother of six ran home and told the mother: "Peter's gone with the soldiers." And Peter's mother cried, but not very long, for she was too busy raising food for four other children to bother about one.

Marin was very useful in the army. Four other boys from

the village also went with him; there were in all fifteen youngsters with a thousand soldiers. The boys were never given rifles or revolvers, but they did the most dangerous work of scouting. When the soldiers drew near to the enemy and did not know in what village he lay in ambush, then Marin and the other boys went to look the ground over.

In their worn linen suits they wandered into the villages, played with the village boys, mixed with the crowds in the markets and listened to what was going on. Russia had many wandering orphans asking for bread, and no one paid much attention to them. Then in the evening Marin would slip out of the village and make his way back to the army and report what he had learned. When the time for fighting came, the soldiers shooed the boys out of the way.

"Get to the rear, kids," they ordered. "There's going to be shooting." Marin wanted very much to see the shooting, but he knew that orders are orders in an army, and he always obeyed.

So Marin doesn't think he really saw the war at all, or was in any danger. When I ask him if it was dangerous being a scout, he laughs at the idea. "Dangerous going into villages like the one he was born in? What danger could be in that?" "But if the enemy had known who you were and what you were doing?" "Oh, then of course they would have shot me," he says. "But how could they ever find out? I was away and on to the next village. I never went back to the same village twice."

After eleven months the fighting was over in the district where Marin served. He went back home to his mother and she scolded him for running away and asked why he had gone. "Oh, just to see and know," said Marin. "My father, too, died for the Revolution."

And now, at home, Marin began working in earnest. He was thirteen and oldest man in a family of six. The land of the lord who once owned the village was now empty and unploughed for the Revolution had driven the lord away and taken his land.

So Marin's mother went to the Land Committee and told them that she was the widow of a soldier who died for the Revolution, and that even her little boy of twelve had fought in the Red Army, and she asked for land for her family. They gave her ten acres.

On these ten acres Marin ploughed and raised grain. It was very heavy work for a boy of thirteen and he worked far beyond his strength. In the fall of the year, lifting many times the heavy wooden fork loaded with grain, he at last strained his wrist. Even now when he bends it a certain way, you can hear it crack.

But all this heavy work did not save Marin and his family. For the Hungry Year came next to the Volga. With it came cholera in the summer, and Marin's mother caught it. "It finished her off quick in two weeks," says Marin. "For cholera is a proper disease that kills quick and sure." And now the food began to fail in the house where Marin lived with his five brothers and sisters. For Marin's work on the land was not as good as a grown man's, and so his harvest was even smaller than the other peasants'. There was so much hunger in Marin's village that people ate other people. But there was a young peasant living near them who had food enough for himself and one more person. He wanted Marin's older sister for his wife, so she went to live with him. And after that the rest of the children were put in a children's home in the town of Kvalinsk.

Two years he lived in the children's home and in all that time he never tasted butter or sugar. There was only black sour bread and not enough of that. But Marin learned to read and write in the home, and that was something. Then he was sixteen and they turned him out and told him he must shift for himself now, for there were many younger children who needed to be fed.

Back Marin went to his home across the river. But now he could no longer farm his land, since he had neither horse nor cow. He went to work for a peasant who wanted an extra farm hand. All year he worked for the peasant but he never got any wages. Just some more black sour bread and a dirty second-hand shirt.

But for all that Marin liked it better than in the children's home.

"For I wanted to work," he tells me. "I can't sit without work. My head turns round and I get dizzy if there isn't any work to do. It was dreadful in the children's home, just sitting and sitting. They taught you reading and writing, but that was all; there weren't any workshops. There wasn't anything you could learn to do."

At last one day, Marin came to Kvalinsk on a market day to sell and buy goods for the peasant he worked for. There in the market he met Nazipaef and Putoff, two boys he had known in the children's home. "What are you doing now?" they asked him. And Marin said he was working for a peasant.

"We are in a colony of other boys, working the land. There we are living quite well now from our own work, and are learning machinery," they told him. "Why don't you come with us?"

So Marin went to the leader of the colony and said he wanted to join. And Yeremeef told him: "Write out an application of what you want to do." Marin wrote that he was sixteen years old and his father was a mechanic who died for the Revolution, and he wanted to learn machines. There was a general assembly of the children and they voted to take Marin in. In the winter he worked in the mill, as his father once did before him, and when summer came and a new tractor arrived at Alexeivka, Marin was already so good at machinery that they chose him to learn the tractor, together with two other boys.

Marin is used to hard knocks and hard work. He takes them all grinning. The other day he had an accident on the tractor. It caught in some heavy weeds and the bottom part stopped, but the top part went right on going and turned over, carrying Marin with it. He managed to jump just in time to save his life, but his head was cut clear open to the skull with an ugly gash mixed with dirt. He tied a handkerchief round it and hiked four miles down to Alexeivka to the little hospital there. They washed the wound and tied it up, and told him he must go to the town of Kvalinsk for proper treatment. So Marin came to my room cheerfully grinning and asked me for his boat fare to Kvalinsk.

To go to Kvalinsk you must wait at the dock all night, for the boat is due at midnight, but it is likely to come any time before morning. Marin, with his bandaged head, lay curled up on the dock till four in the morning, and then rode on the boat till six, and then walked seven miles to the town of Kvalinsk and waited till the hospital clinic was open and it came his turn to be cared for. They sewed up his wound properly, and he walked back to the dock again and spent another night on the way back. He reached Alexeivka at dawn and walked at once to the fields to work. That was the way Marin always acted about any wounds or hurts.

In all the general assemblies Marin always demands that the older boys like him shall have more work than the younger ones. There is always a big discussion on this; sometimes it is settled one way, sometimes another. "I can work all day from sun-up to dark," boasts Marin, "and it won't hurt me now. But when I was little, I sprained my wrist from working too hard and it will always be a bad wrist. And our younger ones now are working beyond their strength; I see them with heads and backs aching in the hot sun. Then they come back and drink lots of cold Volga water and get sick with malaria. There should be two standards of work, for the older ones and the younger ones."

But the younger boys always argue against Marin, for they will not admit that he can do any more work than they can.

Marin and ten of the older boys want to organize their own grown-up commune and have land of their own and farm it together. Marin explained the idea to me one evening. "We are eighteen years old; why should the state help support us like children? But we do not wish to leave the colony one by one to earn our living. We wish to organize a group together. And since we have no horses or cows or machines, let us live for a year

63

more in the colony, keeping accounts of our own work. If we use the horses and tractor so many days, we will pay for it after harvest. Then from the land that we work we will soon have horses and a tractor of our own, and shall build a house and some day get married."

But so far Yeremeef will not let Marin and the others work as a separate group. He says he needs them to teach the younger boys to work; he cannot let all the older ones separate off at once. So Marin has settled down now to work in the mill till spring. "I must be a real machinist," he says, "and learn how to repair all sorts of tractors and machines as well as run them. Then perhaps we can start our grown-up commune here, or I will go to Moscow or maybe even to America and see all the kinds of machines there are."

# VIII. Twelve Doctors Sat on Shubina

WHEN Shubina lay ill for months in the hospital at Kvalinsk, and twelve visiting doctors from Saratov examined her and found that she had pneumonia and bronchitis, and that almost everything else inside her was out of order, I could have told them what ailed her. Shubina was too unselfish; she had a bad case of devotion to other people's welfare. And since that is a very rare disease, none of the doctors recognized it.

Shubina was not used to attracting so much attention. A quiet, retiring girl, she was used only to working and working for everybody without notice. She had done it all her life, ever since she can remember. She tells me with a smile how important she felt when the twelve doctors put her in the sunlight and used magnifying glasses on her to see what was the matter. She herself thinks the matter was that all her body had been frozen. But the real trouble began long before that.

When the pioneer group that took Alexeivka was battling with the coming winter, cooking and eating from one large pail for twenty boys and girls, ploughing, repairing windows, cleaning, while October winds changed into the storms of November and the blizzards of December--Shubina was one of the most energetic, healthiest girls. Nothing was too hard for her. She washed doors with water from the frozen river; she mended the boys clothes by candlelight; she cooked in the single pail in which also water was brought; and she still managed to keep enough vitality to take the only woman's part in the drama which they got up to make money.

The colony sent her as delegate to the party meetings in Alexeivka. And Shubina was very proud of the colony and always stood up for it. Once in the meetings she heard them blaming the colony and saying how badly they kept order, and how they had meat now, and potatoes and bread and even oil and honey--everything you could ask, but the girls were too stupid to cook it.

The other delegates from the colony sat still and hung their heads, for they did not dare talk back to grown-up people. But Shubina wouldn't stand it. She got right up in the meeting. "How can we cook?" she said, "when we have no dishes. You peasants have only three or four in your families to cook for, and yet you have every one of you a couple of iron pots. But we have nearly thirty to cook for, and we have just one iron pail. We get water in it, and we milk the cow in it, and we cook the soup in it, and wash the dishes in it. We have to drink up all the milk, or else leave it in the soup. And we have to eat all the soup, or else throw it out before we can wash the dishes. And we cannot have any water to drink till after the dishwater is thrown out. Can any of your wives do good cooking like that?"

After that nobody talked again against the colony in the meetings, but Shubina went to the leader of the colony and said: "We won't cook any more without pots, for they all make fun of us for living like pigs." So he sold some wheat and bought two iron pots and a pan. Then sheet-iron came from town, and the boys made pails and a water-tank in their blacksmith shop. And life began to get better in the colony, partly because Shubina fought for it.

Shubina was seventeen years old, but she was just learning to read and write. She had never had a chance before. And now, the Young Communists of the colony decided to give a play and raise money for paper and pencils. Shubina took the part of a girl in "Springtime Without Sun," a play of student life before the Revolution. She was to be a priest's daughter who secretly helped the revolutionists.

The play was a great success. The young actors traveled to the Cherumshan Houses to perform. They also gave it in the village of Alexeivka, and won renown, so that the peasant girls no longer disdained to go out walking with the ragged boys of the colony. For these boys were now "artists" and could stalk proudly through the village market and return the nods of the people who had clapped them in the play. Then the actors decided to give a show in the village of Selidba.

A blizzard came up from the Volga on the cold day in late January when they set forth. It grew cold, cold, colder; the horses staggered through the snow. Shubina had no coat of her own, but she borrowed a thin cotton one from one of the girls. She had no stockings and only a pair of broken sandals on her feet. She covered herself as well as she could with straw. All the children huddled together under the straw, for none of them was any too warmly clad. Four hours they traveled thus to the village of Selidba, and when they arrived, every boy had frost-bitten fingers or toes or ears. But Shubina's arms and legs and face were black with frost; all the last part of the way they had to pound her to keep her from falling into the frozen sleep of death.

When they reached the village school house in Selidba, they hauled Shubina out of the wagon and pounded her some more, till the blood came back aching and painful into her arms and legs. Then she went on the stage and played the part of the student girl who helped the Revolution. The audience of peasants applauded loudly, but it was very poor in money. When they took up the collection there was just 70 cents received for the show.

Shubina spent the night on the stove in a peasant hut and traveled back next day. The blizzard was over; the sun shone on the snow; and although she shivered much in her thin cotton coat and her head was aching, she did not turn black again. But as soon as she reached Alexeivka there came a call from the stables. Odintsof was going to Volsk with horses; did Shubina want to go?

A pillow and two dresses were in Volsk belonging to Shubina; there was also three dollars due her from the time when she worked as a nurse-girl. And the papers about her birth and family were in Volsk, and Shubina needed them. She sprang into the other sleigh and set off with Odintsof.

All afternoon till long after sundown they traveled the fifty miles to Volsk over the snow, while Shubina crouched in the straw in her cotton coat and broken sandals. She got the three dollars and the dresses and pillow and the papers about her birth, and came back next day through the cold, windy weather. By this time her head was hot and aching and her body was shivering cold.

Next morning the carpenter boys came to say that the Big House was at last repaired. All the windows were made and put in, and the walls were whitewashed. But the floors were piled with pieces of plaster and paint and straw and wood. It must be cleaned and scrubbed and that was the work of the girls. And all of the girls refused. For there was no heat in the Big House, which had stood all winter in the cold winds from the Volga. Snowdrifts lay even in the rooms. The water for washing came from holes in the ice. And the girls were barefoot, with three pair of shoes between them.

"We will not clean the Big House until you heat it," they said. And everyone knew that there was not enough wood for heating.

Then Shubina spoke up, arising from the bed where she had lain down with an aching head. "Come on, girls," she cried. She took up a pail of water in which pieces of ice were floating, and a cloth that was frozen stiff, and she put on her leaky sandals and started across the half mile of snow to the Big House. The other girls, ashamed to lag behind, followed her, shivering.

All day long Shubina washed floors and windows in the Big House. Her sandals were laid aside, as they were no protection against the wet, and her bare feet were soaked again and again with the icy water on the floors. Her head grew hotter and hotter and her eyes glazed till she could hardly see. At last she staggered and nearly fell downstairs. Then the girls said: "Go home, Shubina."

Shubina started home across the snow. But on the way she remembered that she was cow-girl, and the cows had not been milked. So she took the pail and went to the cold barn and milked the cows. She fed the calves and the three little pigs. She just managed to reach the house after that and fell unconscious on her bed. They sent at once for the doctor and he came that evening. Shubina already was raving in delirium with a temperature of 104 degrees.

The doctor said she had pneumonia and bronchitis, and that there seemed to be something wrong with her heart and her stomach. None of her internal organs appeared to be working. She was too sick to be taken away to the big hospital in Kvalinsk, and there was no room in the little hospital of Alexeivka, so Shubina lay in the little room with the six other girls. She grew thinner and thinner for it seemed there was nothing she could eat. Shubina says that all her insides had been frozen.

Shubina lay in the little room while the spring thaw warmed the air, and the snow ran off the land through little ravines to the ice of the river, till it, too, melted and moved sluggishly downwards and the great Volga was free again. She lay while the season of mud came and went and the ploughing began. Then at last the roads were firm and the air was warm, and they moved her to the big hospital in Kvalinsk just in time for the big commission of doctors to see when they came up the river from Saratov inspecting the hospitals of the province.

The doctors put her in the sun and looked at her through glasses. They made tests of blood. They felt her all over. Then they said that everything inside of Shubina was out of order and that probably she would die. Perhaps she might live if she had only very soft eggs to eat. But there were no eggs in the colony, and the hospital had no extra money for feeding people, so Shubina lay without eggs or anything else but boiled water: Till at last the American Miss Craves came and bought eggs for

Shubina. "I think without her I should have died," says Shubina now.

It was blazing midsummer when Shubina came back to the colony. Work was in full swing in the fields of Alexeivka, but the days were heavy and stifling, and in the long, white evenings clouds of mosquitoes arose from the sloughs near the Volga and came through the unscreened windows. If you closed the windows at night, you fainted from the closeness of the air; if you opened them you tossed all night fighting mosquitoes. Malaria arose and spread with the clouds of mosquitoes along the Volga shores. Into such a place came Shubina, weak from illness, with orders to rest completely for six months and eat only white bread and eggs, and milk.

Neither white bread nor eggs were in the colony. Shubina had some milk and heavy black bread of rye which turned her stomach and threw her again often on the bed with fever. Malaria took her, and she shook with chills and burned with heat. But there was plenty of quinine from Moscow, and Shubina took it, except sometimes when the druggist was too busy to mix the doses and then for a day or two everyone would be sick again.

And still, between hours of lying on her bed, Shubina kept on working. The doctor scolded her and threatened her with death. Yeremeef, who scolded almost everyone else for being too lazy, would almost curse Shubina back to her bed when she tried to work at the thresher. But Shubina couldn't stand it to sit idle when others were working.

She too, like Stesha and Morosof, thought always about order in the colony. There was so much to be done, and sometimes the others refused to do it. "It is not our turn to work," they would say. But always, when Shubina saw things that needed doing, she did them. And there was too much for one sick girl to stand.

They gave her the job of cow-girl, for that was supposed to be easy. She rose at four, milked the three cows, fed the calves and pigs, strained the milk and took it to the kitchen. Then she was through till evening. Everyone wanted Shubina for cow-girl because she always did things on time. Besides, some of the other girls, when they were cow-girl, would drink up more than their share of milk. But Shubina, even when she could eat no other food, never took more than her share.

There came a terrible Sunday when two of the girls in the kitchen fell ill with malaria lust at lunch time and had to be helped home to bed. Two others took their places, but by Monday noon these also were in bed with fever. There was only one girl left to work in the kitchen. But two girls who were perfectly well sat idly in their rooms, eating tomatoes which they had stolen from the colony's vines. "It is not our turn to work in the Kitchen," they said. "We worked last week." It wasn't Shubina's turn either, but that made no difference. "I cannot leave Vera alone," she said, and she added the kitchen job to the job of cow-girl.

One evening late, when supper and milking were over, Shubina told me the tale of her life and how she came to the colony, and what she wanted to do in the future. For sixteen years she had wandered from place to place, uncared for, or cared for only by strangers.

"My mother died when I was two years old," she told me, "and my father was very poor, a peasant not far from Volsk. So he gave me away to a boat captain who worked on the Volga and who had no children of his own. I half remember that it was very nice with them on the Volga steamboat; they loved me and were kind. But they died when I was six, and the first thing I really remember was the big funeral they had in Saratov.

"Then I lived with my mother's mother, but she also was poor and pave me away to be nurse-girl. These people were also kind; they treated me like a daughter and gave me not only food but clothes. Once when I was living with them I saw my father; he went by in a cart and they said: 'That is your father.' But already he was past, and he did not know me. I did not try to see him again. For he gave me away when I was little; he is no more to me now than any stranger.

"For many years I worked as nurse girl and servant girl. The first kind people went away to Nijni and did not take me with them. Sometimes my masters were kind and gave me dresses; sometimes they only gave me food. Then I cut up my mother's dresses and made them over; she was a peasant and her dresses were of strong hand-woven linen. They lasted a long time; but now they are all gone. I had also a beautiful silk shawl, as large as a tablecloth; this my mother got one Christmas when she worked for Lord Vorontsef. I sold it for six roubles and bought cotton goods to make a coat.

"When the Revolution came, I was working for a rich peasant. They had 200 acres of land and thirteen cows, and they ate all kinds of things in their dining-room. But to me they gave only a little saucer of thin cabbage soup with black bread. They were the worst people I ever worked for; I was always hungry with them. The workers who served them were also hungry. Once the old man's wife was sorry for the workers and took them some milk to put in their soup. But the old man came back from the fields and caught her at it, and he went into a rage and threw the soup plates in her face.

"I was just eleven years old, and I had to haul water from the well. The well was deep and the bucket was heavy and it often pulled me off my feet. I was afraid of the well; once I nearly fell in it when the bucket pulled me. When I think of how I lived in those old days, I begin to cry. I cannot stand it. In the colony now there is much disorder because of the lazy ones. But at least we are all equal and we have a chance to learn.

"From the time of the Revolution I began to want to learn. I could not read and write, but that was the way with all servant girls. None of us thought of learning to read before the Revolution. But now the czar was gone and the stingy old peasant

I worked for said that was very bad. But the Communists said it was good, and that we must finish with all oppressors. And I thought: I also am one of the oppressed ones and I am tired of it. I will join the Young Communists and learn to read and write and not be oppressed any longer.

"So from the very beginning I tried to join the Comsomol in Volsk. There was a girl I knew who belonged; she had studied in a regular school. She gave me a book and began to teach me to read. But the old man and his wife forbade me to go; they hated the Comsomol. Then I went to visit friends, and sneaked off afterwards to the meeting. This also they discovered and began to reproach me. So I went to bed early, and then climbed out the window and went to the Comsomol. But their little boy found reading-books in my room and told his mother; and after that they fired me out of their house.

"I worked in so many places that I could not go to the meetings often. So I never could be a real member, but still they let me work for them. They put me on committees for collecting money for homeless children. I couldn't yet read, so I could not do much work for the Comsosol. But everything I could I did. And they gave me books and I went on learning to read.

"But I grew tired of working in different places. I wanted to be in one place where I could learn. So I went to a children's home and they sent me here. It is better here than anywhere else I have been. They have let me join the Comsomol and they sent me as a delegate to the Village Party Meeting. They let me act in the plays. And if only we have order this winter, and a school and a sewing-shop, then at last I shall learn a trade and begin to be something.

"But if not--if there isn't any wood, and the school doesn't open, and we have to study in bed as we did last winter; if the sewing machines get smashed and the tailor is too lazy to mend them; and the lamps get broken and no one can make order--then I can't live through another winter. I think I shall die here.

"If everyone works hard, we can make here a good life. But some do not work and do not care. And it is hard to work when you have no shoes; it is hard to wash clothes when there is not enough soap. Even when there is enough food, some of the girls spoil it because they don't know how. It is hard to make order in the kitchen, when the girls fall sick while getting lunch. But yet, somehow, this must be done, if we are to live at all; and still more, if we are to learn. And if we are to make something in the world, better than the old life that is gone."

# IX. The Musical Shoemaker

MOST children came to the John Reed Colony because they were sent by children's homes or boards of education; and some drifted in like tramps and stayed for the sake of food and shelter. But Morosof came because he believed in the ideal of the colony and wanted to give his life to it. He was sixteen years old when he applied for entrance--a young man making his own decisions.

Morosof is leader of our orchestra now; he is also one of our best shoe-makers. He is always being elected chairman of responsible committees. We trusted him to go to Saratov for a five days' buying trip to select mandolins and guitars and balalaikas, and to purchase all the tools needed in the shoe-shop. And he brought them back on the steamer, and sat up all day and all night without sleep to guard them. "For I had valuable baggage, and who would look after it if I went to sleep?" he said.

Wherever Morosof goes, his pleasant serious manner makes friends for the colony. Down in Saratov he met the head of the State Trust that was trying to take away the big Alexeivka farm from the children. After he saw Morosof he changed his mind completely. "I thought you had only bad street boys who would steal the apples of our orchard," he said in amazement. "I never dreamed you had big, responsible fellows like this..." Then Morosof grew red and redder and said not a word, for he knew that even the big boys sometimes stole apples. But when he came back to the colony, he told the boys about it, and said they must all behave now, since the head of the Trust believed in them.

Morosof helps organize our group of shoemakers. In another year, he thinks, if they have tools and a chance to practise, they can let the instructor go and take orders all by themselves. They can work for peasants and even perhaps for shoe stores. They can take in younger boys and teach them also, and they can get books and learn the theory of shoe-making as well as the practice. And books about leather and where it comes from.

Then, according to Morosof's ideas, he will not leave the colony at all, to hunt for a job elsewhere, but the colony will become a grown-up commune without any hired instructors. Only, remembering how the state helped them get started when they were homeless children, they also will take in ever more and more homeless children, growing always, taking more farms, and mills and creameries and factories, till they spread like a great commune into all parts of the state of Saratov, and become part of the organizing life of the greater growing Commune of the United Soviet Republics of the World.

That is what Morosof saw in the John Reed Colony, when it was only a badly organized band of hungry children, sleeping in haystacks or on floors of ruined buildings, and when he himself was an orphan in the village of Selidba, and a member of the Young Communists there. Now he is president of our Organizing Committee, and a member of the Bureau of our Comsomol, besides being our chief musician.

Morosof's father was an unskilled worker on the docks of Astrakhan, and his mother was a servant girl and a laundress, working in hotels. Most of our children come from large families, and have three to ten brothers and sisters; but Morosof was an only child, for his father died in the year of his birth. His mother lived by odd jobs in lodging houses and private families, and young Morosof was knocked about in servants' quarters until he was ten years old. It is hard to see how this life produced our president and chief musician.

The Revolution came to the city of Astrakhan when Morosof was ten years old. He remembers the strike of the soldiers and the disorder that followed. Hungry soldiers went around robbing stores, and underfed Persian workers revolted against their Russian masters and went out with long knives to take what they could find. Always there were meetings and voting and discussions: "Who is for the czar? Who is for the Soviets?" The manager of the hotel where Morosof's mother worked went round among the servants and paid them five or ten roubles apiece to vote for the czar; but this was only a few cents in the already sinking money. For all that, the vote went heavily for the Soviet power.

Young Morosof hung round meetings when the Red Guards organized and elected officers, and unearthed or commandeered weapons, and declared that they would establish "revolutionary order" and stop the lawlessness and stealing. He saw the taking of the fortress in Astrakhan, and the fighting which sent up in flames all the buildings around the Theater Square. Then, when the boss of the hotel fled away, together with the other bosses, Morosof's mother, who felt strange and insecure without a master, went home to her almost forgotten folks in the northern village of Selidba. "She was not very developed and did not understand the Revolution," says young Morosof, excusing his mother. He himself was eleven years old then, but he understood it.

Typhus raged across the land in the path of retreating, demoralized, ragged soldiers. The district hospital in Selidba was filled to over-flowing, and Morosof's mother got easily a job as laundress. It was a fine job, with high wages; but it was deadly to Morosof's mother. She caught typhus from the soiled linen and died a few weeks later, leaving her boy with his poor and aged grandmother.

And now, for the first time, Morosof went to school, and finished three classes in two years. Then came the Hungry Year and he went with five hundred other children in a freight car to the lands of bread. Two winters he spent in Kostroma and finished the fifth grade there. He studied natural science and geography, and mathematics and hygiene and the Russian language. He is the best educated boy in the whole colony. He

77

also studied shoe-making. Then they sent him home to Selidba, for he was sixteen years old now, and he lived again with his grandmother and worked for peasants in return for food.

It was there that he heard of the colony, and dreamed his dream of what it might be, and came and applied to the General Assembly of the children, and he was voted in. "It is better here than in the children's homes," he said to me. "For here we own the colony; there is no limit. They do not say: 'When you are sixteen, you must leave and make room for others.' But the older you grow the better you can work for the colony. It is all ours and we must all work to make it better. That is how life should be in a commune.

"When I entered the shoe-shop there was no leather, and only instruments enough for six people. Then Yavorskaia came from Moscow bringing sole-leather, and Poliakof came with instruments. But to get leather for uppers we had to wait often, buying ten roubles worth at a time. Yet we made 140 pair of shoes last winter for the colony, and six pair for the teachers and five pair to order for peasants. We also repaired all the 140 pair at least twice. We would have enough shoes for everyone now, but they sent forty boys from Volsk, who were hooligans picked up from the streets. They stayed just long enough to get shoes and then they went away, taking their own shoes and another pair with them. They also stole blankets and sold them in the Volsk bazaar. It is hard to make order, and so much work goes for nothing.

"We have fourteen shoemakers now. Eight of them are good ones. Four can make shoes even without an instructor, good enough shoes to sell in the market. I think we should not take any more shoemakers yet, because there is not leather enough. And we all need practice, practice. If we take more shoemakers, no one will have practice and we will all be long in learning.

"But if we fourteen have plenty of practice this winter, then we can organize our shop without any instructor, and begin

78

to look for orders and even work for the market and the shoe-stores. Then we can take in more and more young shoemakers and train them also, teaching them ourselves without any paid instructor."

I asked Morosof what he wanted to do when he became an expert shoemaker. Would he go to the city to look for work? "I want," he said, "to help build the commune. It is as well here as anywhere. I have no home, and no property, and the city does not interest me. I have been to Saratov and to Astrakhan; they are both big cities, but I like it better here. We will build our commune here, and no one will say: 'This is mine, and that is yours'--but everything we see, all the land and the buildings and the mill and the shops, we will say: 'They are all ours.'

"But Kersoff has gone off to Baku, and I thought he was truly one of us. He went because he had an uncle and a chance to be his own boss and own property. But I do not see the use of property; I think it is better not to own it. The October Revolution taught us to organize the commune. Even peasants begin now to do this; how much more can homeless children, who have no homes or property to begin with?"

"But," I inquire, "when the colony grows up large, you cannot have more than two hundred people. And if you marry and have families, there will be no room; and if you take in more homeless children, there will be still less room. But if you stop taking in homeless children, then you will be just a grown-up group of two hundred people, living for yourselves and with property belonging to two hundred people instead of to one. How is that any better?"

Morosof had his answer ready. "There are many Soviet farms badly run, and losing money. We can take them over; we can have new sections to our colony. And after the Soviet farms are gone, there is still much empty land, when we shall be strong enough to build our own houses and barns. And if the homeless children shall all be gone, there are still children of peasants, who live very badly and will want to join our commune. And the land will not be gone, and the children will not stop coming--no, not in all the time that I shall live.

"But the bad thing is that our organization is weak. There is Kersoff, who was a good comrade, and who goes off to Baku to have property. There are others who are lazy and dirty and who eat the tomatoes off the vines instead of putting them in the kitchen. There are girls who want the job of cow-girl so that they can drink more than their share of milk. Half of the colonists are Young Communists, but many are not much developed. And Volsk sends us boys from the streets, who steal our shoes and blankets.

"Yet already we see that life goes better than last year. Last year we worked one hundred and fifty acres in four different fields; this year we are working nearly four hundred in one big farm. Last year there were tools for six shoemakers, this year for fourteen. Last year we had no school, but now we have repaired a school, with club-rooms and a theater. We are repairing the big mill and it will give us much bread. We have repaired the thresher and are threshing grain for all the peasants, and earning food from it also. We are even going to have an orchestra, guitars, mandolins, balalaikas!

"But most of the serious ones are sick, because we did the work of the others. Nazipaef, Stesha, Shubina and I--we are all sick often. For when the harvest came and the malaria with it, it was too much for us. But now in winter is the time of rest and learning. We have both food and fuel. We must strengthen our organization and enforce its will on the shirkers. The shoemakers must take orders and earn money to buy tools for a better shoe-shop. We must have order and keep accounts, so that the shoe-making money does not all go to buy cows, leaving the shoe-shop to stop from wornout tools. This happened to our carpenter shop, that last winter took orders. And they made more than a hundred roubles, but the colony bought a cow with it, and now the carpenter's tools are worn out and they cannot take more orders.

Of course we also need cows, and every department must help the whole colony. But it must not ruin itself to do it. For this we need so much order and so many accounts and so much planning."

When Morosof finished speaking, he picked up the shoe-making tools he had bought in Saratov and took them to deposit with the boy who kept the warehouse, receiving a receipt which he took to the office. Then he consulted about the place to put the musical instruments, which every child so thirsted to lay hands on that there was no safe place in the colony. They decided at last to take them to the Soviet Farm and leave them with the orchestra instructor, but he let Morosof himself take care of the mandolin, and practice on it to get ready for the "music circle."

By this time it was nearly six o'clock in the evening and Morosof had spent two days and a night without sleeping. But still without rest, he set off for the sixteen-mile walk to the Cherumshan Houses to deliver an important letter to Yeremeef.

And I went out to the kitchen, and thence to the new school and club-house where the whitewashing is almost finished. I saw the groves near the Big House filled with human animal filth from our disorderly habits; I heard boys complaining from the kitchen that the girls had baked sour bread; and the girls complaining that the tractorist would not give them kerosene for their lamp when they had to bake at night, and that the boys had stolen the lamp anyway; I heard one girl weeping because a boy had torn her last undershirt as it hung on the picket fence to dry; and the boy retorting that she had torn it in the wash... But I also saw Shubina, arisen from malaria, trudging off to the barn to milk the cows at the appointed hour.

I saw gardens filled with weeds, broken fences, disorder. But beyond them I saw acres of ploughed fields, black and soft, through which Marin, our tractorist, was steadily pushing the tractor. Black and rich they rolled up to the horizon and there met other acres of stubble from which we had just cut a whole year's

81

supply of wheat. And down from a gap in the hills came great hayricks loaded With lucerne from our farthest acres, where the land lies in a gentle valley close to the pine and oak forests. And against the western sunset rose the smoke of our donkey-engine, where the thresher was threshing grain for a dozen peasants.

The cloudless blue of the sky darkened slowly, and filled with a million stars. And the noisy wrangling in the kitchen was drowned by the songs of the boys returning from the thresher and the lucerne fields, weary and glad of the work accomplished. And I thought of Morosof, still trudging on for another three hours in the starlight, to carry his important letter to the Cherumshan Houses. And I knew that in all his talk of work, and accounts, and organization, and sickness--of lack of tools, and lack of faithful workers, yet steady, painful progress--he had been telling me the story not only of John Reed Children's Colony, but of the whole great structure of Russia.